An Introduction to
Electrical Instrumentation

An Introduction to
Electrical Instrumentation

A guide to the use, selection, and limitations of electrical instruments
and measuring systems

B. A. Gregory

*Senior Lecturer Specialising in Electrical Instrumentation, Department of
Electrical and Electronic Engineering, Brighton Polytechnic*

Distributed in the United States by
CRANE, RUSSAK & COMPANY, INC.
347 Madison Avenue
New York, New York 10017

First published 1973 *by*
THE MACMILLAN PRESS LTD
London and Basingstoke
Associated companies in New York Dublin
Melbourne Johannesburg and Madras

SBN 333 14833 9

*Text set in 10/12 pt. IBM Press Roman, printed by photolithography,
and bound in Great Britain at The Pitman Press, Bath*

Contents

Preface

Our ability to measure a quantity determines our knowledge of that quantity, and since the measuring of electrical quantities — or other parameters in terms of electrical quantities — is involved in an ever expanding circle of occupations of contemporary life, it is essential for the practising engineer to have a thorough knowledge of electrical instrumentation and measurement systems. This is especially so since in addition to his own requirements, he may be called upon to advise others who have no electrical knowledge at all.

This book is primarily intended to assist the student following an electrical or electronic engineering degree course to adopt a practical approach to his measurement problems. It will also be of use to the engineer or technician, who now finds himself involved with measurements in terms of volts, ampères, ohms, watts, etc., and faced with an ever increasing variety of instruments from a simple pointer instrument to a complex data logging and processing system. Thus, the object of this book is to help the engineer, or instrument user, to select the right form of instrument for an application, and then analyse the performance of the competitive instruments from the various manufacturers in order to obtain the optimum instrument performance for each measurement situation.

During that period of my career when I was employed in the research department of an industrial organisation I was, at times, appalled by the lack of ability exhibited by some graduates in selecting a suitable, let alone the best, instrument to perform quite basic measurements. Since entering the field of higher education to lecture in electrical measurements and instrumentation, my philosophy has been to instruct students to consider each measurement situation on its merits and then select the best instrument for that particular set of circumstances. Such an approach must of course include descriptions of types of instruments, and be presented so that the student understands the functioning and limitations of each instrument in order to be able to make the optimum selection. Hence the general theme of this book is to describe various types of instrument and then compare their characteristics. Unfortunately there is a limit to the number of instruments that can be described in a book of realistic size, and I have therefore omitted

specialised topics such as medical instrumentation, gas chromatography, radio
frequency measurements, power system measurements, acoustic measurements,
and high voltage instruments such as discharge detectors. Further, since some of
the established methods are extensively covered elsewhere they are only ·
summarised in this book; there is a list of references for further reading at the
end of each chapter.

I would like to thank all the instrument manufacturers who have willingly
assisted me in producing this volume by providing application notes, specifications,
reproductions of articles, and also their obliging field engineers. I have endeavoured
to acknowledge all sources of diagrams and other material, but I hope that any
oversights will be excused. I should also like to thank my colleagues in the
Department of Electrical and Electronic Engineering at Brighton Polytechnic for
their assistance and encouragement; in particular my thanks are due to Dr B. H.
Venning and Dr E. M. Freeman. Finally may I record my thanks and appreciation
to my wife for her perseverance and courage in typing the manuscript.

B. A. Gregory

Introduction

Scientific and technical instruments have been defined as devices used in observing, measuring, controlling, computing or communicating. Additionally the same volume* states that: 'Instruments and instrument systems refine, extend or supplement human facilities and abilities to sense, perceive, communicate, remember, calculate or reason'.

The principal concern of this book is to describe instruments capable of measuring and recording the magnitudes and variations in electrical and mechanical quantities, and to illustrate methods by which it should be possible to select the optimum instrument for any measurement situation. With this latter point in mind, and before describing any 'hardware' it is desirable to consider some of the factors that govern the choice of an instrument for a particular measurement situation.

Accuracy

The term 'accuracy' is really one of conversation and in a measurement situation the operator should be concerned with defining the limits of error that apply to a particular measurement. As an example consider a voltmeter that has a fiducial value† of 1 V and is a class 1.5 instrument. Then if this meter is connected across a pair of terminals between which there is a voltage such that the meter indicates 1 V, the 'conventional true value' of the voltage (providing the meter is of a suitable type) is somewhere between 0.985 and 1.015 of an absolute or standard volt. It should be noted that the error quoted is in terms of the fiducial value and the tolerance on a reading of 0.1 V on the same instrument would still be ± 0.015 V or ± 15 per cent of the reading if the measurement is performed at reference conditions†. This aspect is covered in more detail in chapter 7.

* *Encyclopedia of Science and Technology*, McGraw-Hill, 1971.
† see B.S. 89: Part 1:1970

Bandwidth

The bandwidth of an instrument relates to the maximum range of frequency over which it is suitable for use and is normally quoted in terms of 3 dB points. For example, an oscilloscope amplifier may have a specification quoting -3 dB points of 20 Hz and 50 MHz, indicating that at these frequencies the gain of the amplifier will be $1/\sqrt{2}$ or 0.707 of its midfrequency value (that is 29 per cent less, also there will be a 45° phase shift) so that measurements made at or near these values of frequency will have considerable errors.

Screening

Measurements made involving electrical quantities of small magnitude can be seriously affected by electromagnetic and electrostatic interference from external sources. Protection and screening from these effects must therefore be included in a sensitive measurement system. Chapter 4 gives the basic procedures which should be adopted to reduce these 'interference' problems.

Impedance effects

Almost any instrument when connected into a circuit will change the conditions that existed in the circuit prior to its inclusion. It is obviously important to ensure that this disturbance is a minimum or incorrect results will be produced. As a simple example consider a source of d.c. voltage which has an internal resistance of 100 Ω and an open circuit voltage of 2 V (see Figure I.1).

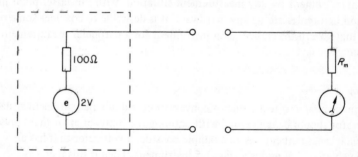

Figure I.1

Connecting across these terminals a voltmeter with an impedance (or resistance) of 100 Ω, a current of 2/200 or 0.01 A will flow and the voltage indicated by the meter will be $2-100 \times 0.01$ or 1.0 V. However if a meter of 1900 Ω resistance is used, the current flowing will be 2/2000 or 0.001 A and the indicated voltage becomes $(2-100 \times 0.001)$ or 1.9 V, which is a considerable improvement. Using a meter with say, a 10 000 Ω resistance would, of course, reduce the error in measurement still further.

A similar illustration of the effects of ammeter resistance is of interest, that is suppose the above source is passing current through a 10 Ω resistance and the current is measured with ammeters of 10, 1, 0.1 Ω resistance; the readings obtained would be 0.01667 A, 0.01802 A and 0.01816 A respectively; the correct value, or the value with no ammeter, being 0.01818 A. If alternating voltages are involved the impedance effects may not be purely resistive and phase angle errors may also be present.

Sensitivity

An instrument's sensitivity will be quoted as so many units for full scale, for example 1 A f.s.d., or as a unit of deflection, for example, 10 mm/μA. It should be noted that in order to obtain the least error in a particular measurement the instrument used must be such that the indicated value is near the full scale value of the most suitable range.

Display method

The display or recording method used in a measurement is usually governed by the purpose of the measurement, hence one must ask, is a pointer or analog instrument's deflection suitable for the application, or is the display in numerical form from a digital instrument more satisfactory? Alternatively, it may be that a permanent record is required, when the choice is from a graphical presentation, a printed output, punched paper tape, or a record on magnetic tape for later analysis.

Waveform purity

In a.c. measurements the quantity which is usually of greatest interest is the root mean square (r.m.s.) value or, expressed in another manner, its equivalent direct current magnitude. For a single frequency sinusoidal wave this presents little problem as the relationships between the peak value, the rectified average value and the r.m.s. value are constants. However, for waveforms with harmonic content these constant relationships are not valid and instruments which actually measure r.m.s. as opposed to those calibrated in r.m.s. quantities must be used.

From this preamble, which only indicates some of the influencing factors, it should be apparent that when making a measurement, it is not sufficient to pick up the nearest meter calibrated in the appropriate units and expect it to yield satisfactory results.

Thus the reason for the ever increasing number of different instruments to perform each type of measurement becomes a little clearer. A further source of confusion is in the use of symbols and units. This book uses the SI system (Système International d'Unités), and to assist the student a list of symbols, units and conversion factors between imperial, MKS and SI units is presented in appendix I.

Analog (Pointer) Instruments

Definition

An analog device is one in which the output or display is continuously variable in time and bears a fixed relationship to the input.

The use of analog instruments is very extensive and whilst digital instruments are ever increasing in number and applications, the areas common to both types are, at present, fairly limited and it is therefore likely that analog devices will remain in extensive use for a good many years, and for some applications are unlikely ever to be replaced by digital devices. Analog instruments may be divided into three groups: (a) electromechanical instruments (Section A); (b) electronic instruments (Section B) which, broadly speaking, are constructed by the addition of electronic circuits to electromagnetic indicators thus increasing their sensitivity and input impedance; and (c) electromechanical and electronic instruments having a modified display arrangement so that a graphical trace, that is a display of instantaneous values against time, is obtained (see chapter 2).

SECTION A. ELECTROMECHANICAL INSTRUMENTS

When an electric current flows along a conductor, the conductor becomes surrounded by a magnetic field. This property is used in electromechanical instruments to obtain the deflection of a pointer: (a) by the interaction of the magnetic field around a coil with a permanent magnet; (b) between ferro-magnetic vanes in the coil's magnetic field; or (c) through the interaction of the magnetic fields produced by a number of coils.

Constraining these forces to form a turning movement a deflecting torque $= G \cdot f(i)$ newton metres (Nm) is obtained which is a function of the current in the instrument's coil and the geometry and type of coil system. To obtain a stable display it is necessary to equate the deflection torque with an opposing or control torque. The magnitude of this control torque must increase with the angular deflection of the pointer and this is arranged by using spiral springs or a ribbon suspension so that the control torque $= C \times \theta$ N m, where θ is the angular deflection in radians and C is the control constant in newton metres per radian and will depend on the material and geometry of the control device[1].

The moving parts of the instrument will have a moment of inertia (J) and when a change in the magnitude of deflection takes place an accelerating torque $(J.d^2\theta/dt^2$ N m) will be present. As the movable parts are attached to a control spring they combine to form a mass–spring system and in order to prevent excessive oscillations when the magnitude of the electrical input is changed, a damping torque $(D.d\theta/dt$ N m) must be provided that will only act if the movable parts are in motion. The method by which this damping torque is applied may be:

(a) eddy current — where currents induced in a conducting sheet attached to the movement produce a magnetic field opposing any change in position

(b) pneumatic — in this method a vane is attached to the instrument move- ment, and the resistance of the surrounding air to the motion of the vane provides the required damping. Fluid damping is an extension of this principle, a small vane then being constrained to move in a container filled with a suitably viscous fluid (see page 52).

(c) electromagnetic — movement of a coil in a magnetic field produces a current in the coil which opposes the deflecting current and slows the response of the instrument. The magnitude of the opposing current will be dependent on the resistance of the circuit to which the instrument is connected.

Combining the above torques, the equation of motion for a pointer instrument becomes:

$$J.\frac{d^2\theta}{dt^2} + D.\frac{d\theta}{dt} + C.\ \theta = G.f(i) \tag{1.1}$$

which will have a steady state solution[3],

$$C.\theta = G.f(i) \tag{1.2}$$

and a dynamic or transient solution of the form (see appendix II),

$$\theta = A.e^{\lambda_1} + B.e^{\lambda_2} \tag{1.3}$$

where A and B are arbitrary constants and

$$\lambda_1 = \frac{-D}{2J} + \left(\frac{D^2}{4J^2} - \frac{C}{J}\right)^{\frac{1}{2}} \tag{1.4}$$

$$\text{and}\ \ \lambda_2 = \frac{-D}{2J} - \left(\frac{D^2}{4J^2} - \frac{C}{J}\right)^{\frac{1}{2}} \tag{1.5}$$

For a particular instrument C and J are fixed in magnitude during manufacture, but D (the amount of damping) may be varied. This results in three possible modes of response to a transient:

(a)   when $D^2/4J^2 > C/J$ – for which the roots λ_1 and λ_2 are real and unequal, and is known as the overdamped case, curve (a) in figure 1.1.

(b)   when $D^2/4J^2 = C/J$ – for which the root are real and equal, and D has a value termed the critical value, curve (b) in figure 1.1.

(c)   when $D^2/4J^2 < C/J$ – which gives roots that are conjugate-complex quantities and the system is underdamped, curve (c) in figure 1.1. The frequency of the decaying oscillations being:

$$\omega = \left(\frac{C}{J} - \frac{D^2}{4J^2}\right)^{\frac{1}{2}} \tag{1.6}$$

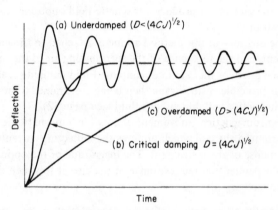

Figure 1.1. The effects of damping magnitude on the movement response

Thus it is apparent that the magnitude of the damping applied to a movement has an important effect on the dynamic performance of an instrument; it being general practice to operate pointer instruments with slightly less than critical damping to ensure that the pointer changes rapidly from one position to another with the minimum chance of sticking.

1.1 MOVING COIL INSTRUMENT

The principle used in the construction of this type of instrument is that current passing through a conductor generates a magnetic field around the conductor and if this field is arranged to interact with a permanent magnetic field, a force acts on the current carrying conductor. If the conductor (that is the coil) is constrained to move in a rotary manner an angular deflection or movement proportional to the current may be obtained, resulting in an instrument that has a linear scale but which, due to its inertia, can only respond to steady state and slowly varying quantities. The linearity of scale is an extremely useful feature and accounts for the use of moving coil instruments as the display in many complex electronic instruments.

The general arrangement of the moving coil instrument is indicated in figure 1.2. The permanent magnet system has over the years been considerably reduced in size due to the improvements in magnet design as better materials have become available.

The coil may be air cored or mounted on a metal former; if the latter is

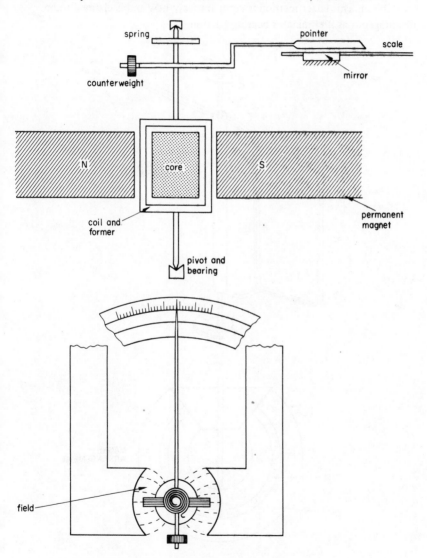

Figure 1.2. Moving coil instrument

present it will contribute to the damping of the instrument (eddy currents) but will add to the inertia of the movement.

If an air cored coil offset from the axis of rotation (figure 1.3) is used, the scale length of the instrument can be increased from 120° to 240° or even 300° enabling a better resolution of reading for the same instrument range.

The suspension may be either a variation on the jewelled bearing 'clock spring' arrangement (figure 1.2) or a ribbon suspension (figure 1.3) in which there are no bearings and where the control torque is derived from the twisting of the suspension ribbon. This latter method is comparatively new and is claimed to be advantageous as it eliminates bearing friction.

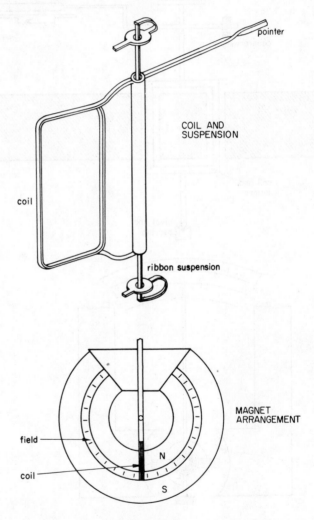

Figure 1.3. 'Long scale' moving coil instrument

The pointer is of lightweight construction and, apart from those used in some inexpensive instruments, will have the section over the scale twisted to form a fine blade, this helping to reduce parallax errors in the reading of the scale; in many instruments such errors may be further reduced by careful alignment of the pointer blade and its reflection in the mirror adjacent to the scale. The weight of the pointer is normally counterbalanced by weights situated diametrically opposite and rigidly connected to it.

To damp the movement a combination of electrical eddy current effects, the passage of the movement plus vanes through the surrounding air, and of friction in the bearings is used.

Features

(a)   Suitable for d.c. only.
(b)   Low power consumption (compared with other nonelectronic analog instruments).
(c)   High torque/weight ratio.
(d)   Linear scale which may be long with an offset coil system.
(e)   Freedom from hysteresis errors, and fairly insensitive to the effects of external fields (its own magnet will shield the coil from these).
(f)   Good damping by eddy currents in the metal former.
(g)   Wide range of sensitivities available.

Applications

As an ammeter

This is a direct application of the moving coil instrument since its deflection is directly proportional to current, or perhaps more strictly the deflection is proportional to the ampere turns of the coil winding. Thus, by using a large number of turns a small current will be sufficient to obtain full scale deflection

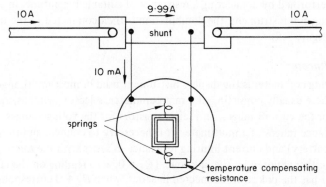

Figure 1.4. Current measurement by moving coil instrument (see also figure 5.10)

– instruments with a f.s.d. of 50 μA are not uncommon – or alternatively, fewer turns and a larger current may be used. The magnitude of this current is limited by the size of the coil conductors which can be used and by the problem of connecting the coil, usually via the suspension, to the external circuit. To measure currents greater than 1 A it is usually necessary to use current shunts, these items are covered in greater detail in chapter 5, but in principle they provide a low resistance path parallel to the instrument coil as in figure 1.4.

As a voltmeter

As stated above the moving coil instrument has a deflection directly proportional to current, and to use this instrument as a voltmeter merely requires the unknown voltage to be converted to a small (milliamp) current proportional to its magnitude, and for this current to be passed through the instrument coil. This operation is

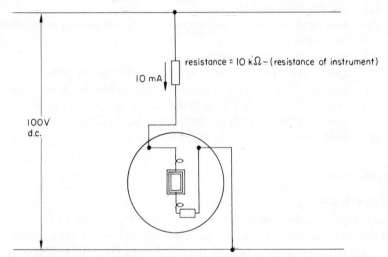

Figure 1.5. Voltage measurement by moving coil instrument

simply performed by connecting a resistance of suitable magnitude, in series with the moving coil instrument; the combination is then connected to the unknown voltage (figure 1.5).

As a multimeter

The moving coil meter is the display instrument used in most multirange instruments, these usually consisting of a microammeter, selector switches, appropriate shunts for the current ranges, and series resistors for the voltage ranges. For the resistance ranges of a multimeter, it is necessary to connect an internal power source (battery) and current limiting resistance in series with the microammeter which is either shunted by the unknown ($R_x = 0$, zero reading on the current range); or has the unknown connected in series, when $R_x = 0$ corresponds to full scale on the current ranges.

As a galvanometer

The simplest form of galvanometer is a direct adaptation of the moving coil ammeter, in which the zero position is at the centre of the scale and the pointer is free to move in a positive or negative direction depending on the current direction in the coil. The use of the instrument is in determining balance or zero current

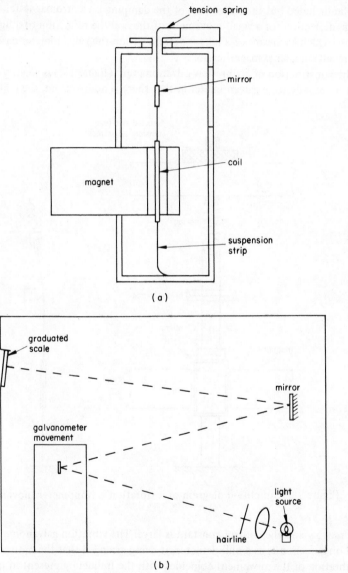

(a)

(b)

Figure 1.6. Light spot moving coil galvanometer. (a) Galvanometer movement; (b) optical system

flow between parts of a circuit. For the measurement of very small currents and for use as a sensitive null detector, galvanometers of high sensitivity (typically 430 mm/μA) are required, when the coil will have a large number of turns (bonded together for maximum strength and stability) and suspended by a high tensile alloy strip, which provides the small control torque and also acts as connections to and from the coil (figure 1.6). A small amount of fluid damping may be included but the major part of the damping is electromagnetic. To obtain a large deflection for a small movement of the coil the reflection of a light spot is used, which by means of a mirror system within the galvanometer case (figure 1.6) results in a large magnification.

The construction of a vibration galvanometer[2] (figure 1.7) is slightly different from the steady state galvanometer in that there is no iron core, the coil is

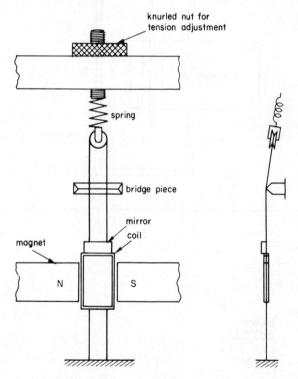

Figure 1.7. Simplified diagram of a vibration galvanometer movement

very narrow and the control constant is large. The vibration galvanometer is used on a.c. circuits as a null detector, it being arranged that the natural frequency of vibration of the movement coincides with the frequency present in the a.c. circuit. The current I, supplied to the coil is $I_m \sin \omega t$ and the amplitude A of the resulting vibration

$$= \frac{G I_{\mathrm{m}}}{[D^2\omega^2 + (C - J\omega^2)^2]^{\frac{1}{2}}} \qquad (1.7)$$

(See appendix IIb)

The frequency at which mechanical resonance occurs may be adjusted either by the knurled nut or by the position of the bridge piece. However, increasing the tension to increase the natural frequency causes a corresponding decrease in sensitivity and it is therefore rarely possible to use the same galvanometer over a wide frequency range. The amplitude against frequency curve is sharply peaked (if the damping is small) so a vibration galvanometer tuned to the fundamental frequency of the a.c. circuit will have little or no response to harmonics in the electrical signal. Vibration galvanometers may be constructed to have resonànt frequencies up to 1000 Hz, but are now normally only used at power-line frequencies.

1.2 MOVING IRON INSTRUMENT

The deflecting torque for this type of instrument is obtained either by attraction (figure 1.8a) of a soft iron vane into the field of the coil, or by magnetic repulsion between two bars or vanes positioned within a coil. The repulsion form

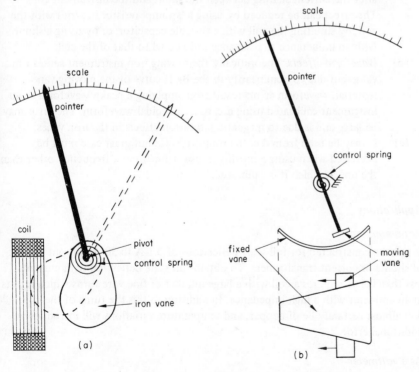

Figure 1.8. Types of moving iron instrument. (a) Attraction, (b) repulsion

of movement is the more widely used since, if shaped vanes are used (figure 1.8b), an instrument with 240° of deflection may be obtained. For either the attraction or the repulsion version of the moving iron instrument it may be shown[2] that the deflecting torque is proportional to $G . I^2$ and it can therefore be used in either a.c. or d.c. circuits. In most instruments the control torque is derived from a helical spring and the damping torque pneumatically.

Features

(a)   Nonlinear scale, but a comparatively robust and inexpensive instrument to manufacture.

(b)   *Indication hysteresis.* The moving iron instrument gives a low indication when measuring a direct current of slowly increasing magnitude, but yields a high reading when measuring a slowly decreasing direct current. This effect is due to the magnetic hysteresis curve of the iron used in the vanes.

(c)   *Frequency effects.* If the range of a moving iron ammeter were increased by using a resistive current shunt, an error would result due to changes in the coil reactance with frequency, since this impedance variation will alter the current sharing between the shunt and the instrument coil. The effect can be reduced by using a 'swamp' resistor in series with the coil; by shunting the coil with a suitable capacitor; or by using a shunt with an inductance to resistance ratio equal to that of the coil.

(d)   *Waveform effects.* Theoretically the moving iron instrument senses r.m.s. value but due to nonlinearity in the B–H curve of the moving vane material, waveform errors result from applying a peaky waveform to an instrument calibrated using d.c. or a sinusoidal waveform. The error may be large and is due to magnetic saturation effects in the iron vanes.

(e)   It may be used from d.c. to around 100 Hz but great care must be exercised when using a moving iron instrument at a frequency other than the one at which it is calibrated.

Applications

As an ammeter

It may be constructed for full scale deflections of 0.1 A to 30 A without the use of shunts or current transformers. To obtain full scale deflection with currents less than 0.1 A requires a coil with a large number of fine wire turns which results in an ammeter with a high impedance. In addition to this the turns of the coil will almost certainly be of copper, and temperature variations will introduce an additional error.

As a voltmeter

The moving iron voltmeter is a fairly low impedance instrument, typically 50 Ω/V for a 100 V instrument, and the lowest full scale is of the order of 50 V, it being impracticable to manufacture these instruments with lower full scale values and a low current consumption. In addition to this, the major part of the applied total voltage will occur across the instrument's coil, and this being made of copper will produce an instrument with a large temperature coefficient.

Power factor meter[2]

The measurement of the phase angle between voltage and current in an a.c. system is of importance and an instrument which measures this angle directly is the moving-iron power factor meter, in which the angular position of the moving-iron vanes is determined by the phase angle of the line current with respect to the phase voltages.

1.3 THE ELECTRODYNAMIC INSTRUMENT

This instrument, often termed a dynamometer, relies for its deflecting torque on the interaction of the magnetic fields produced by a pair of fixed air cored coils and a third air cored coil capable of angular movement and suspended within the fixed coils (figure 1.9). The deflecting torque produced within the instrument is proportional to the product of the current in the moving coil (I_m) and that in the fixed coil (I_f). that is torque $\propto I_m \cdot I_f$.

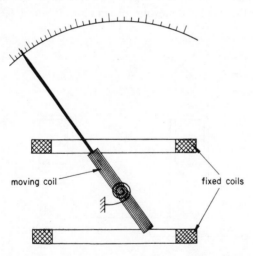

Figure 1.9. Coil arrangement of the electrodynamic instrument

Note: this is an approximation, since the torque is also dependent on the initial angle between the axes of the coils[12].

This instrument is suitable for a.c. and d.c. measurement of current, voltage and power. As no iron is present within the coil system the magnetic fields are weak and to obtain sufficient deflecting torque requires coils of either a large number of turns or capable of carrying a moderate current. The control torque is derived from a helical spring and the movement damped by air resistance.

Features

(a) Voltmeters and ammeters have approximately square law scales; watt-meters have approximately linear scales[1].

(b) It is more expensive to manufacture than either the moving coil or moving iron type. It has a higher power consumption than the moving-coil type.

(c) It measures true r.m.s. values of a.c. waveform irrespective of waveshape.

(d) It is an instrument suitable for use in either a.c. or d.c. circuits, being comparatively unaffected by frequency. Some models are available for use over the range d.c. to 2.5 kHz[5].

(e) Stray magnetic fields can affect the operation of the dynamometer and therefore the coil system is enclosed in a magnetic shield (see page 145).

Applications

As an ammeter

For the measurement of small currents (5–100 mA) the instrument would be arranged to have the fixed and moving coils in series (figure 1.10). With this type

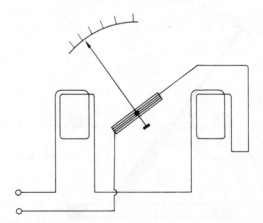

Figure 1.10. Coil connections for electrodynamic milliammeter

of arrangement the dynamometer milliammeter can conveniently be used as a transfer instrument (see page 83).

For measuring larger currents (up to 20 A) it becomes impracticable to pass the full value of current through the moving coil, so this is then shunted by a

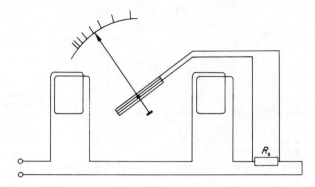

Figure 1.11. Coil connections for electrodynamic ammeter

low resistance (figure 1.11). This arrangement could be used for still larger currents, but in most cases it is better to use a current transformer (see page 163) and an ammeter with a full scale deflection of 5 A.

As a voltmeter

Since full scale deflection may be obtained with as small a current as 5 mA, the electrodynamometer can be used for voltage measurement by connecting a non-reactive resistance of low temperature coefficient in series with the instrument coils. The minimum full scale deflection for such an instrument is generally 30 V and the maximum 750 V.

As a wattmeter

This is the widest use of the electrodynamic instrument. Since its deflection is proportional to the product of the two currents, it is usual to make the current in the moving coil proportional to the circuit voltage and to make the fixed coils carry the load current (figure 1.12). Depending on the relative magnitudes of the currents in the fixed and moving coils the wattmeter should be connected as in figure 1.13a or figure 1.13b to obtain the minimum error in the measurement of power in a load. In figure 1.13a the current coil will produce a force proportional to the load current only, but the voltage coil (moving) will produce a force proportional to the voltage across the load plus the current coil. However, in figure 1.13b the current coil (fixed) produces a force proportional to the sum of the currents in the load and in the voltage coil, while the voltage coil produces a force proportional to the voltage across the load.

In either case correction for the error may be made but the wattmeter should always be connected to indicate with minimum error.

Let I_w be the current through the wattmeter current coil; I_L be the current in the load; V_w the voltage across the wattmeter voltage coil and V_L the voltage across the load.

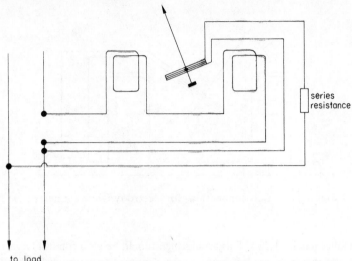

to load

Figure 1.12. Electrodynamic wattmeter, internal connections

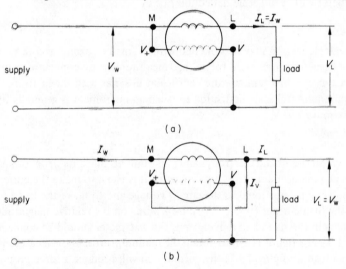

Figure 1.13. Wattmeter connection conventions. (a) High impedance load
(b) low impedance load

Let R_c and R_v be the resistances of the wattmeter current and voltage coils respectively, and R_L the load resistance.

Additionally, let jX_c be the inductive reactance of the wattmeter current coil and jX_L the inductive reactance of the load.

The inductive reactance of the voltage coil may generally be ignored since it will be 'swamped' by a series volt-dropping resistor.

For the connection in figure 1.13a.

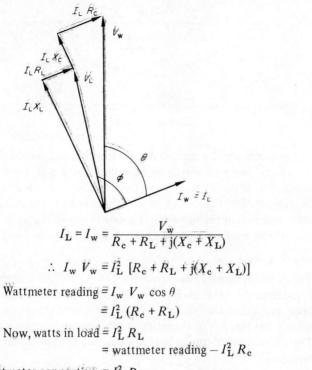

$$I_L = I_w = \frac{V_w}{R_c + R_L + j(X_c + X_L)}$$

$$\therefore I_w V_w = I_L^2 [R_c + R_L + j(X_c + X_L)]$$

Wattmeter reading $\equiv I_w V_w \cos \theta$

$$\equiv I_L^2 (R_c + R_L)$$

Now, watts in load $= I_L^2 R_L$

$$= \text{wattmeter reading} - I_L^2 R_c$$

hence the wattmeter connection $= I_L^2 R_c$

(1.8)

For the connection in figure 1.13b

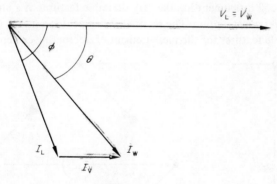

Wattmeter reading $= V_w I_w \cos \theta$

$$= V_w (I_L \cos \phi + I_v)$$

$$= V_L I_L \cos \phi + V_L I_L$$

$$= V_L I_L \cos \phi + \frac{V_L^2}{R_v}$$

hence the wattmeter correction $= \dfrac{V_L^2}{R_v}$

(1.9)

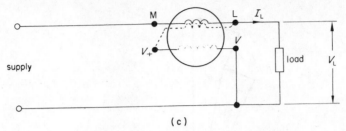

(c)

Figure 1.14. Compensated wattmeter

The corrections for a.c. and d.c. are identical and **should always be applied**.

In a compensated wattmeter[12] the above error is eliminated by an internal connection of the V+ terminal to the L terminal, by a lead which is wound into the current coils, figure 1.14, so that the current to the voltage coil cancels out the proportion of the fixed coil flux due to the voltage coil current passing through the fixed coils when V+ is connected to L.

Power factor meter

The electrodynamic power factor meter has two coils mounted on the shaft at approximately $90°$ to each other and connected to line voltage one via a resistance and to the other via a reactance. The angle between the coils[2] is slightly less than $90°$ if the series impedances are resistance and capacitance. There is no control torque in such an instrument but pneumatic damping will be present.

1.4 RECTIFIER INSTRUMENT

The moving coil instrument has the very desirable features of a uniform scale and low power consumption. The rectifier instrument is a means of adapting these desirable features for the measurement of alternating current.

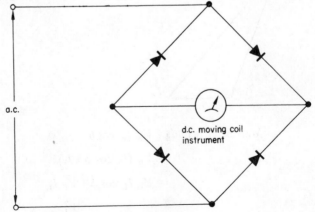

Figure 1.15. Full wave rectifier instrument (see also figure 5.26)

The adaptation is performed quite simply, and without modifying the construction of the movement, by placing a copper oxide or silicon rectifier bridge between the a.c. circuit and the d.c. moving-coil meter[4], figure 1.15, which will then indicate the rectified or 'average' value of the a.c.

The heating effect of an alternating current, that is its r.m.s. value, is 1.11 times the 'average' or rectified value (providing the waveform is a single frequency sinusoid), and rectifier instruments which always sense the average value are most calibrated in terms of r.m.s. values.

Features

(a) A.C. only; typically 20 Hz to 10 kHz.

(b) Calibration assumes pure sine waves, the presence of harmonics gives erroneous readings.

(c) The majority of the scale is approximately linear, but the low end is nonlinear due to the rectifier characteristics (see page 176).

(d) The rectifier is temperature sensitive, and errors can therefore result from changes in the ambient temperature.

Applications

As an ammeter

The rectifier instrument is not particularly suitable for use in measuring current but it may be used as a microammeter and low milliammeter (up to 10 or 15 mA). For larger currents the rectifier becomes too bulky and shunting is impracticable due to the rectifier characteristics.

As a voltmeter

The most suitable range of operation for a rectifier instrument is in the medium (50–250 V) voltage range. Higher impedances than for either the moving iron or the dynamometer voltmeter are normal, and the frequency range is wider, although an instrument calibrated at power frequencies should not be expected to retain its calibration for audio frequency measurements.

As a multimeter

This form of a.c. measurement is the one normally used for the a.c. ranges of multimeters. The voltage ranges are obtained by the use of suitable noninductive series resistors and the current ranges by use of a current transformer (see page 163).

1.5 THERMOCOUPLE INSTRUMENTS

These instruments also employ a technique which enables the moving coil instrument to be used for the measurement of a.c. quantities. The conversion from a.c. to d.c. is in this case performed by using the alternating current to heat a small element, the temperature of which is converted to a direct current by a thermocouple attached to it. A thermocouple circuit is the term applied to two lengths of dissimilar electric conductor, joined at the ends to form a closed loop (see page 267). If the junctions of the dissimilar metals are maintained at different temperatures a current that may be measured by a sensitive moving coil instrument flows in the loop. Since this current is approximately proportional to the temperature difference between the hot and cold junctions, and if the cold junction is maintained at a constant temperature, the current in the moving coil instrument will be proportional to the temperature of the heater, which in turn is dependent on the r.m.s. or heating effect of the current in the heater. In many instruments of this type the terminals of the moving coil instrument will form the cold junction, figure 1.16, and will be subject to the variations in

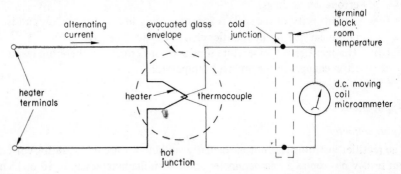

Figure 1.16. Principle of the thermocouple instrument (see also figure 8.44)

ambient temperature. However, these variations are generally insignificant (say 10–25 °C) compared with the temperatures of the heater which may be of the order of 1000 °C for the heater current corresponding to full scale deflection.

It should be noted that if a thermocouple is being used to measure the temperature of an object it is often more satisfactory to open circuit the dissimilar metal loop and measure, using an instrument with a high imput impedance, the voltage that appears across the open circuit.

Features

(a)   Measures true r.m.s. value of an alternating current irrespective of its waveshape.

(b)   Wide frequency range — may be used from d.c. up to the megahertz range, but pointer vibrations may be experienced at low frequencies (less than 10 Hz).

(c)   Fragile, having a low overload capacity. Normally exceeding a 50 per cent overload will melt the heater element.

(d)   May have a higher impedance than moving-iron or electrodynamic instruments.

(e)   Nonlinear scale; the temperature of the heater will be proportional to current squared, also the characteristic of a thermocouple is slightly nonlinear.

Applications

As an ammeter

Mainly limited to the milliampere ranges (2–50 mA f.s.d.) but may be constructed for larger currents. It is however one of the most suitable methods for measuring currents at radio frequencies.

As a voltmeter

Since the instrument is satisfactory as a sensitive ammeter, it merely requires the addition of a series resistance for use as a voltmeter. However, if it is intended that the instrument shall be used at high frequencies, it is important for the series resistance to be 'pure' (see page 123), that is, nonreactive, or its impedance, and hence the current in the heater will change with frequency.

As an a.c. to d.c. converter

The thermocouple instrument is one of the few methods of determining the true r.m.s. value of an a.c. waveform. For precise measurement, the heater/hot junction and the cold junction are situated in an evacuated container and the d.c. output measured by a good quality digital voltmeter or potentiometer (see pages 77 and 197). By this technique the a.c. to d.c. conversion may be performed with an error of less than 0.05 per cent over a range of frequencies from 20 Hz to

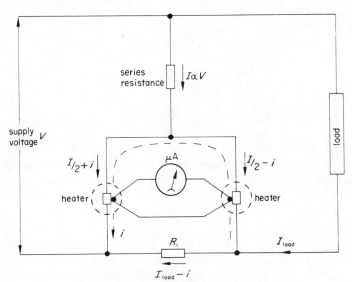

Figure 1.17. Thermocouple wattmeter

1 MHz; it also applies for higher frequencies (50 MHz), but with reduced accuracy. Reference 6 describes the construction of a thermal converter with a sensitivity of 1 part in 10^6.

As a wattmeter

For measuring power at frequencies up to 1 MHz a form of thermocouple instrument[4] may be used. A suitable arrangement is shown in figure 1.17, where a current heating effect proportional to the voltage is applied to two thermocouple heaters so that the thermocouples produce equal voltages on no load. When load current passes through the resistance R_1, a small circulating current will add to the current in one heater and subtract from the other, causing a temperature difference between the thermocouples and a reading on the microammeter.

1.6 ELECTROSTATIC INSTRUMENTS

This type of instrument depends on the forces set up between charged plates to obtain a deflecting torque. The total force is proportional to the product of the charges on the plates and since the charge on a conducting plate is proportional to voltage the deflecting torque of an electrostatic instrument is proportional to

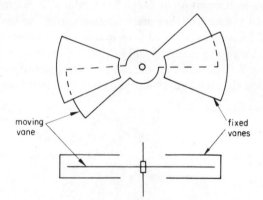

Figure 1.18. Electrostatic voltmeter vane arrangement

V^2. The deflecting torque is derived from the force acting on a moving vane that is suspended in fixed quadrant-shaped boxes, as in figure 1.18. The control torque is derived from a spring system and damping is usually pneumatic.

Features

(a)   Measures true r.m.s. values, because the deflecting torque is proportional to V^2 and can therefore be used as a transfer instrument.

(b)   Used for d.c., and for a.c. where it constitutes a capacitive load, this limiting its general application to frequencies below 100 Hz. However, it may be used to radio frequencies[4] where it could be employed for aerial tuning.

(c)   It is a comparatively fragile and expensive instrument.

(d)   A high input impedance instrument, on d.c. its resistance being that of the leakage path through the insulation.

(e)   Nonlinear scale, approximately square law, but this may be modified by the shape of the vanes.

Applications

As a voltmeter

To obtain sufficient deflecting torque for medium voltage electrostatic instruments (minimum f.s.d. around 120 V), it is necessary to operate a number of vanes or plates in parallel, figure 1.19, the use of a strip suspension eliminating bearing friction.

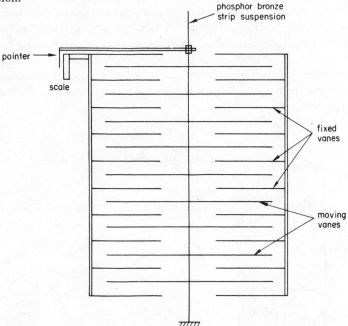

Figure 1.19. Multiple vane, medium voltage electrostatic voltmeter

For higher voltages the gaps between fixed and moving vanes may be increased and the number of vanes reduced.

As a wattmeter

The electrometer type of e.s.v. may be used as a wattmeter[1,4]. In the electrometer instrument there are two sets of fixed plates and a pair of moving vanes, so that applying a voltage difference (proportional to current) between the sets of fixed plates enables this type of electrostatic voltmeter to be used as a wattmeter.

1.7 ENERGY METERS

The amount of electricity used by a consumer must be metered and so energy meters are perhaps one of the most familiar electrical instruments.

The requirement here is for a display or integrated record of the electrical energy used over a period of time. This is performed by causing an aluminium disc to rotate (and hence drive a number of counting dials) at a speed proportional to the product of the voltage supplied to and the current in phase with the voltage taken by the consumer. That is, a given number of watts will have been consumed for each revolution of the disc, and by totalling the number of revolutions the energy consumed will be recorded[2, 10, 11].

To rotate the disc, eddy currents in it (induced by the combined action of a magnetic flux proportional to the supply voltage E_s and a magnetic flux proportional to the consumer's current I_s) produce a magnetic flux opposing the inducing flux, resulting in a torque on the disc proportional to $E_s I_s \cos \phi$ where $\cos \phi$ is the load power factor. It should be noted that although frequency is absent from this expression for the torque on the disc, it will affect the induced eddy currents and hence the torque. Thus the instrument is only suitable to use at its calibrated frequency.

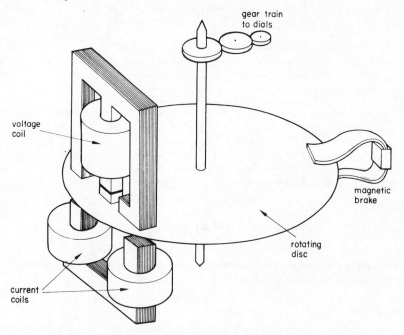

Figure 1.20. Energy meter components

A braking torque must be provided and this is obtained by using a permanent magnet to induce eddy currents of a frequency proportional to the speed of rotation of the disc. This form of instrument is shown diagrammatically in figure 1.20.

SECTION B. ELECTRONIC INSTRUMENTS

The electromechanical instruments, with the exception of the electrostatic instruments, have one undesirable feature in common; namely that if used as a voltmeter they present the measurand (quantity being measured) with a low or medium impedance and the deflection obtained is dependent on the current drawn from the circuit on which the measurement is being performed. In many applications this is of little consequence as the low to medium impedance of the instrument may be many times greater than the source impedance, and the current drain of the instrument small in comparison to the current in the circuit. However, when the source impedance is large, such a measuring device will seriously affect the value measured and may also affect the operation of the circuit under observation.

Placing a large impedance in series with any of the preceding instruments would limit the current to such an extent that the resulting pointer deflection would be minute or nonexistent. However, small signals may be amplified electronically and the resulting larger signal used to deflect a pointer instrument, which is commonly a moving-coil meter.

1.8 D.C. VOLTMETERS

The measurement of d.c. voltage is a straightforward application of the above principle, that is a d.c. amplifier preceding the moving-coil indicating instrument. The type of d.c. amplifier used will largely depend on the sensitivity, freedom from drift, and accuracy desired and will either be of the 'direct coupled' or 'chopper' type.

Direct coupled amplifier

The original valve voltmeter (v.v.) made use of the high impedance between the grid and cathode of a triode, figure 1.21. This technique is still in use in the electrometer type of electronic instrument where a specially constructed valve is incorporated, the input terminal of the instrument being directly connected to the grid via a terminal in the glass envelope and not in the base. This results in a grid–cathode resistance of the order of $10^{17}\,\Omega$; these special valves have a low anode voltage, and a grid current of about 1 per cent of that of a conventional valve.

Another form of circuit is shown in figure 1.22 where the input signal passes through a range selector to a field effect transistor (f.e.t.) (used so that a high impedance shunts the range selector which may have an impedance of 100 MΩ), and then to other transistors which form a direct coupled amplifier of sufficient gain to drive the meter. The input impedance of such an instrument will be high and correction for loading is not often necessary. An additional feature often provided is an output proportional to the meter reading which may be used to

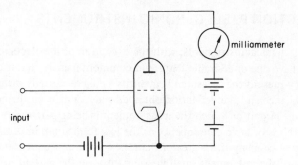

Figure 1.21. Basic valve voltmeter

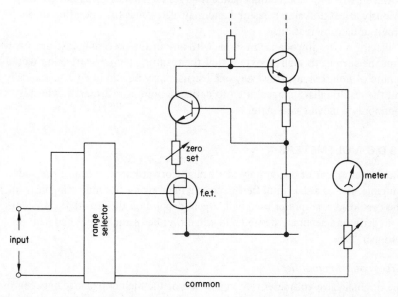

Figure 1.22. Solid state direct coupled voltmeter (Hewlett Packard)

drive a pen recorder or enable the instrument to be used as a narrow band amplifier (see page 43).

Some instruments of this type are battery operated and this is useful if it is desirable to isolate the instrument from the power supply for a particular measurement.

The direct coupled amplifier type of instrument is attractive for its economy and thus finds its major application in the lower cost electronic instruments. It is however limited in sensitivity as it is impracticable to make it drift free; for example, temperature variations over a period will change the amplifier output and if the voltage to be measured is small these variations may be larger than the amplified quantity.

Chopper amplifier type

For very sensitive instruments – full scale a few μV – a chopper amplifier is normally used for the first stage of amplification. In such an amplifier the d.c. voltage is chopped (figure 1.23) to a low frequency (200–300 Hz) a.c.; passed through a blocking capacitor; amplified; passed through another blocking

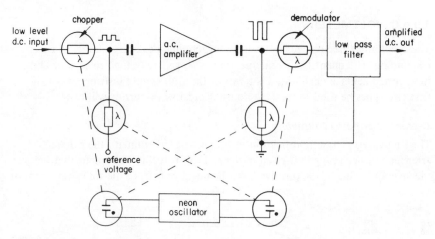

Figure 1.23. Photoconductive chopper amplifier (Hewlett Packard)

capacitor (to remove and d.c. drift or offset from the amplified signal); demodulated or rectified; filtered; and then applied to a moving coil ammeter.

By suitable selection of the decoupling or blocking capacitors it is possible to manufacture a sensitive instrument which is virtually drift free, its overall accuracy and high d.c. input impedance being enhanced by substantial d.c. feedback from the output to the modulator (see page 172).

The basic arrangement within such a device is shown in figure 1.23 which demonstrates the use of a photoconductive (or photoresistive) chopping technique, in which a light sensitive semiconductor component having many megohms resistance when unilluminated changes its resistance to a few hundred ohms when lit by a neon or incandescent bulb. An alternative method of chopping the low level d.c. input is to use an electromagnetic chopper[8], but the frequency of operation of this form is lower, commonly power line frequency, and will not prevent errors due to power line frequency pickup. A commercially available instrument using a photochopper amplifier has an input impedance of 100 MΩ, a resolution of 0.1 μV and input ranges from 3 μV full scale to 1 kV full scale with a stated accuracy of ± 2 per cent f.s.d. It should be noted that the 100 MΩ input impedance is unlikely to apply to all ranges.

A further form of d.c. 'electronic voltmeter' is that which incorporates a self-balancing potentiometer (see page 45). In an instrument called 'maxiscale' produced by George Kent Ltd[13], the difference voltage between the input signal and the voltage on the potentiometer is amplified and applied to a motor that is used to drive the scale and potentiometer slide wire past a pointer and slide wire contact until the voltage on the latter is equal to the input signal. This type of instrument is eminently suitable for indicating small d.c. voltages such as the output from thermocouples when the display could conveniently be in K or °C.

1.9 A.C. VOLTMETERS

These fall into three categories: average responding, peak responding, or r.m.s. responding. The quantity of principal interest is generally the r.m.s. value and most instruments are calibrated is terms of this, but should instruments of the first two types be used to measure nonsinusoidal waves, errors will result[7].

Average responding voltmeters

The first category is a modification of the rectifier instrument, the general arrangement becoming that given in figure 1.24. It is this instrument that measures the 'average' value of the positive half, the negative half, or the total waveform of

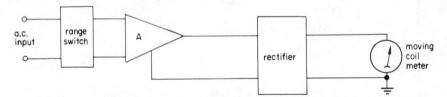

Figure 1.24. 'Average' measuring electronic voltmeter

a cycle, depending on the type of rectification used. It is however usually calibrated in terms of r.m.s. values and for an input waveform with a form factor other than 1.11 erroneous measurement will result.

$$\text{Note:} \qquad \text{form factor} = \frac{\text{r.m.s. value}}{\text{average value}} = 1.11 \text{ for a pure sine wave}$$

The a.c. ranges of an electronic multimeter will in general sense the average value of the waveform and care must be exercised when for example measurements are being made on circuits incorporating silicon controlled rectifiers (see page 21).

Peak responding voltmeters

If the alternating input signal is half-wave rectified and then applied to a capacitor it will be charged to the peak value of the applied voltage. This value of d.c. voltage may then be amplified and used to deflect a moving coil meter (figure 1.25). These instruments can perform over a bandwidth extending to several hundred MHz and have a good linearity for input signals of 0.5 V and above.

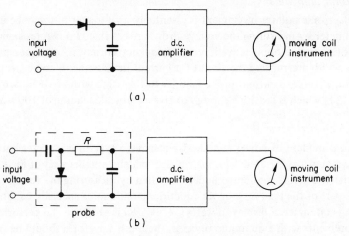

(a)

(b)

Figure 1.25. Peak responding voltmeters

For lower voltages special compensation techniques must be adopted to retain a linear meter scale. However, like the average responding meter, the peak responding instrument is usually calibrated in terms of r.m.s. values (r.m.s. = 0.707 peak for a sine wave), but they are more sensitive to harmonic distortion and care must be exercised in their use.

R.M.S. responding voltmeter

The thermocouple instrument is an r.m.s. sensing instrument and to extend its range to the measurement of small signals only requires in principle the addition of amplification of the input signal prior to its being used to heat the thermo-couple. In practice, to remove thermocouple nonlinearities and obtain a linear

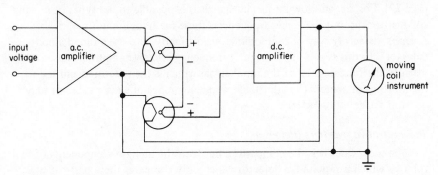

Figure 1.26. True r.m.s. sensing electronic voltmeter

output display, an arrangement of the form given in figure 1.26 may be used. These are r.m.s. measuring devices, unaffected by any amount of wave distortion providing peak excursions do not exceed the dynamic range of the instrument.

Electronic ammeters

The electronic pointer instrument is essentially a voltmeter but may be used as an ammeter by measuring the voltage drop across a fixed known resistance, which must of course be nonreactive if a high frequency current is to be measured.

To avoid interrupting the circuit for current measurements some electronic instruments have a current probe (similar to that used with an oscilloscope, see page 137) which is sensitive enough to give a full scale reading of 100 μA.

Electronic wattmeter[9,14,17]

For measurement of power at audio frequencies a type of wattmeter known as a power meter may be used. This instrument, which contains a set of load resistors, is connected across the source under test, then by measuring the voltage drop across part of the load resistor via a rectifier circuit and a specially calibrated moving-coil meter, a display in watts ($W = V^2/R$) is obtained. To perform measurements with a minimum of error, the input waveform should be sinusoidal, and the impedances of the power meter and the source matched. Such an instrument may have as many as forty values of load impedance which can be individually selected by a front panel switch.

An example of its use is in measuring the output power from an a.f. amplifier into a specific load impedance.

More recent developments are: an electronic wattmeter that utilises a thermocouple arrangement to determine power consumption over frequencies up to 1 MHz for low powers (3 mW) and 250 kHz for 'high' powers (3 kW), and the use of multipliers for determining power consumption.

1.10 NULL DETECTORS

It is often necessary to detect small currents and reduce them to zero (see page 77). For d.c. and low frequency signals it is possible to use types of moving coil instrument, but for higher frequencies these methods are not practicable. A very satisfactory way of overcoming this difficulty is to use a tuned amplifier detector[8]. In this type of instrument a frequency selective amplifier is used to magnify the fundamental of the input signal which is then rectified and fed to the moving coil meter. The adjustable frequency range of such a detector may be from 10 Hz to 100 kHz.

Phase sensitive rectifiers (detectors)

Since in most cases it is the comparison between the inphase components of two voltages which is required, a detector which only compares these parts of two voltages will be advantageous in a large number of cases.

Consider two voltages (figure 1.27) which are of different magnitudes and have a small phase angle between them, it being desired to compare only the component of V_2 in phase with V_1.

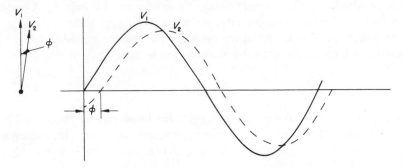

Figure 1.27. Voltages to be compared

Let the voltage V_1 be applied to a transformer whose secondary is centre tapped and connected to the circuit in figure 1.28. In the positive half-cycle of V_1 current will flow round the path XBAQ and the meter would deflect, say, to the right; in the negative half of the cycle current would flow round the path

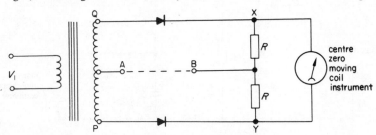

Figure 1.28. Principle of phase sensitive detector (rectifier) (see also figure 5.30)

YBAP the meter deflecting to the left; but, providing the frequency of V_1 was greater than a few hertz, the resulting meter deflection would be zero, that is the true average over one cycle.

Suppose the link AB is now removed and the voltage V_2 is applied between these points. The total voltage driving current around path XBAQ in the positive half-cycle is then the sum of V_1 and V_2 and the meter needle will deflect further to the right. However, during the negative half-cycle when P is negative to A, B is also negative to A, and the resulting voltage driving current around YBAP is $(V_1 - V_2)$ and the meter deflection to the left will have been reduced. Thus the average deflection of the meter over a complete cycle will be to the right of the centre zero, the magnitude of this deflection depending on the relative magnitudes of V_1 and V_2.

If V_2 were antiphase ($180°$ out of phase) with V_1 the average deflection for one cycle would be to the left. Thus a detector is obtained which is sensitive to magnitude and capable of comparing the relative phase of a single voltage with a reference.

The circuit discussed above can be used satisfactorily in practice, an alternative being to interchange the reference voltage V_1 and the signal voltage V_2. A possible addition can be a smoothing capacitor across the moving coil meter.

To reduce the vibration effects on the pointer of the moving coil meter, circuits which yield a full wave rectification are advantageous[18].

1.11 THE Q METER

The Q of a coil is the ratio $\omega_0 L/R$ where L is the inductance of the coil and R is the resistance of the coil at the frequency of measurement ω_0. This resistance R is an involved and frequency dependent quality since its value includes eddy current, and skin effects.

The method used for determining Q is to apply a known variable frequency voltage (E_s) across a series LC circuit (figure 1.29), and to vary the frequency applied to the series circuit until resonance causes E_q to reach a maximum.

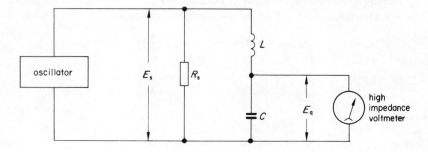

Figure 1.29. Principle of a Q meter

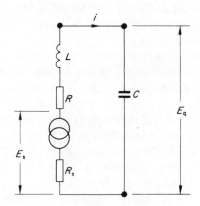

Figure 1.30. Equivalent circuit of a Q meter

The resistance R_s is small (typically 0.02 Ω) so that the resonating circuit is fed from a low impedance source. Thus the equivalent circuit of figure 1.29 becomes that in figure 1.30 and the current

$$i = \frac{E_s}{j\omega L + R + R_s + \dfrac{1}{j\omega C}} \qquad (1.10)$$

This current has a maximum value of

$$i_{res} = \frac{E_s}{R + R_s} \approx \frac{E_s}{R} \text{ (since } R_s \text{ has been made small)}$$

occurring at the resonant frequency ω_0, that is

$$j\omega_0 L + \frac{1}{j\omega_0 C} = 0$$

$$\text{or} \quad \omega_0 = \frac{1}{(LC)^{\frac{1}{2}}} \qquad (1.11)$$

At this resonant frequency the voltage E_q across the tuned circuit is

$$E_q = i_{res} \times \frac{1}{j\omega_0 C} = \frac{E_s}{R} \times \frac{1}{j\omega_0 C} \qquad (1.12)$$

$$\text{Thus} \quad \frac{E_q}{E_s} = \frac{1}{\omega_0 CR} = \frac{\omega_0 L}{R} = Q \qquad (1.13)$$

Note: If the resistance of the inductor is also low, for example 1 Ω the magnitude of R_s will introduce an error in the measurement of Q, that is, apparent Q.

$$Q_{ap} = \frac{\omega L}{R + R_s}, \qquad (1.14)$$

$$\text{and true} \quad Q = Q_{ap}\left(1 + \frac{R_s}{R}\right) \qquad (1.15)$$

The measurements of the voltage across the capacitor must be made using a very high impedance meter such as a vacuum tube voltmeter (v.t.vm.) (or loading of the turned circuit will result) and since the Q of the circuit is also the voltage magnification, the v.t.vm. may be calibrated in terms of Q providing E_s is adjusted to a predetermined value for all readings.

The practical Q meter is a straightforward adaptation of the basic circuit that is shown in figure 1.31.

The measurement of E_s may be performed as a current measurement using a thermocouple ammeter since the effect of the resonant circuit on the current taken from the variable frequency oscillator will be insignificant. The unknown inductance (L_x) is connected between the Hi and Lo terminals; C_q being a variable air dielectric capacitor. Should it be desired to measure an unknown

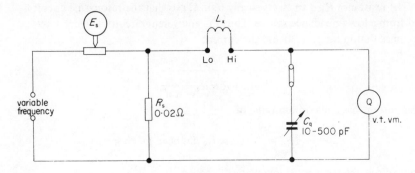

Figure 1.31. Basic Q meter circuit

capacitance instead of an inductance, a known high Q inductor must be connected between the Hi and Lo terminals and the unknown C connected in parallel with C_q between the Hi and E terminals. One commercially available equipment is operable over a range of frequencies from 1 kHz to 300 MHz.

Applications

Determination of inductor properties

As indicated above the unknown inductor is connected in the tuned circuit and either the capacitor C_q or both it and the oscillator frequency are adjusted to obtain resonance, this being indicated by the maximum value of Q for the prescribed level of E_s, when

$$Q = \frac{\omega_0 L}{R}$$

also at this frequency

$$\omega_0 L = \frac{1}{\omega_0 C_q} \tag{1.16}$$

This being the indicated or apparent inductance of the coil at the frequency f_0.

The resistance R at this frequency

$$= \frac{1}{\omega_0 C_q Q} \; \Omega \tag{1.17}$$

It is often necessary to know the distributed self-capacitance (C_0) of an inductor, and a Q meter may be used to determine this as follows:

(a)   setting C_q to the maximum value C_1 (typically 500 pF) a resonant frequency may be obtained. Let this be f_1.

Then

$$f_1 = \frac{1}{2\pi \; [L(C_1 + C_0)]^{\frac{1}{2}}} \tag{1.18}$$

(b) doubling this frequency so that $f_2 = 2f_1$ and adjusting C_q for resonance gives

$$f_2 = \frac{1}{2\pi \, [L(C_2 + C_0)]^{\frac{1}{2}}}$$

Then

$$f_2 = \frac{2}{2\pi \, [L(C_1 + C_0)]^{\frac{1}{2}}}$$

$$\therefore \quad C_0 = \frac{C_1 - 4C_2}{3} \qquad (1.19)$$

and is the self-capacitance of the coil under test.

To obtain the true inductance (L_{true}) as opposed to the indicated or apparent inductance of a coil it is necessary to take this self-capacitance into account as follows:

$$L_{true} = \frac{L_{ind} \cdot C_{ind}}{(C_{ind} + C_0)} \qquad (1.20)$$

Since $\quad L_{true} = \dfrac{1}{(C_{ind} + C_0)} \quad$ and $\quad L_{ind} = \dfrac{1}{C_{ind}}$

The difference between L_{true} and L_{ind} is usually small and in most practical cases may be neglected.

Determination of capacitor properties

The properties of a capacitor that it may be desirable to determine are: capacitance, loss angle, and leakage resistance.

(a) Small values of capacitance ($<$500pF).

The determination of capacitance is straightforward providing it is less than the maximum value of C_q. The measurement is performed by selecting a suitable, high Q, known inductor, connecting it between the Hi and Lo terminals and obtaining a resonance with C_q set to a maximum. Let these values be C_1 (500pF) and Q_1 occurring at a frequency f_1.

Connecting the unknown capacitor in parallel with C_q and adjusting its value to restore resonance will give values of Q_2 and C_2; also at frequency f_1. Then

$$C_x = C_1 - C_2 \quad \text{pF} \qquad (1.21)$$

Treating this condition as one of a parallel circuit, that is, figure 1.32a

$$Q_1 = \frac{\omega_0 L}{R_s} = \frac{R_p}{\omega_0 L_p} = \omega_0 C_p R_p$$

Now

$$Q_2 = \frac{R_p{'}}{\omega_0 L_p} = \omega_0 R_p{'} C_p$$

where
$$R_p' = \frac{R_x R_p}{R_x + R_p}$$

$$\therefore R_x = \frac{R_p R_p'}{R_p - R_p'} \tag{1.22}$$

$$= \left(\frac{Q_1 Q_2}{(Q_1 - Q_2)}\right) \frac{1}{\omega_0 C_p} \tag{1.23}$$

R_x being the leakage resistance of the capacitor C_x now

$$C_p = C_1 + C_0 = C_2 + C_x + C_0$$

(C_0 = stray capacitance of the standard inductor)

$$\therefore R_x = \frac{Q_1 \cdot Q_2}{Q_1 - Q_2} \cdot \frac{1}{\omega_0 (C_1 + C_0)} \tag{1.24}$$

And the loss angle of the capacitor

$$\delta = \tan^{-1} \frac{1}{\omega R_x C_x}$$

$$\therefore \tan \delta = \frac{(Q_1 - Q_2)}{Q_1 \cdot Q_2} \times \frac{(C_1 + C_0)}{(C_1 - C_2)} \tag{1.25}$$

In addition to measuring the properties of a capacitor the above techniques may be used to determine the input impedance of a device or $R-C$ network.

(b)   Large values of capacitance (> 500pF).

This is performed by connecting the unknown capacitance (shunted by a known resistance R_{sh}) in series with C_q, as in figure 1.32b. Again two 'balances'

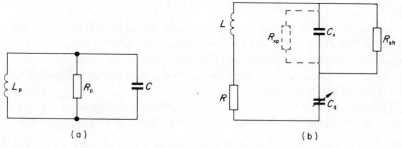

(a)   (b)

Figure 1.32. Equivalent circuits for the evaluation of capacitor properties

must be made. The first with C_x short circuited giving readings of C_1; f_1; and Q_1 for which

$$Q_1 = \frac{\omega_1 L}{R} = \frac{1}{\omega_1 C_1 R} \quad (C_q \text{ having been set to approx. midvalue for } C_1)$$

With C_x in circuit the meter is rebalanced by adjustment of C_q to give Q_2 and C_2 also at f_1 then

$$\frac{1}{C_2} - \frac{1}{C_1} = \frac{1}{C_x} \text{ or } C_x = \frac{C_1 C_2}{C_1 - C_2} \qquad (1.26)$$

and the leakage resistance of C_x is

$$R_x = \frac{Q \cdot R_{sh}}{\omega C_x \cdot R_{sh} - Q} \qquad (1.27)$$

where

$$Q = \frac{Q_1 Q_2 (C_1 - C_2)}{C_1 Q_1 - C_2 Q_2} \qquad (1.28)$$

and if $Q > 10$

$$\tan \delta = \frac{1}{\omega C_x R_{xp}} \qquad (1.29)$$

1.12 HALL EFFECT DEVICES

If a material carries a current in the presence of a transverse magnetic field, a voltage (dependent on the strength of the field, the magnitude of the current and the 'Hall effect' coefficient) will be produced between the edges of the material (figure 1.33). The magnitude of the 'Hall effect coefficient' depends on

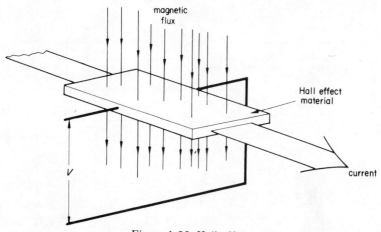

Figure 1.33. Hall effect

the material and in some semiconductors is sufficient to enable the voltage so produced to be used (after amplification) for deflecting a moving coil instrument in a manner that is proportional to the magnetic flux passing through the semiconductor. These instruments are usually called flux or gauss meters and have the following features and applications:

Table 1.1. Comparison Table of Analog (pointer) Instrument Characteristics*

Type	Accuracy class	Sensitivity Max.	Fiducial value[19] Max. current	Fiducial value[19] Min. voltage	Approx. Impedance Ω/V	Scale shape	Frequency Range (Hz)	Remarks
Moving coil	0.1 → 0.5	500 mm/μA	1 A	50 mV	100 → 20 000	Linear	d.c. → ~30	
Moving iron	1 → 5	100 mA f.s.d.	30 A	50 V	50	Non-linear	d.c. → ~100	Cheap
Electrodynamic instrument	0.2 → 0.5	5 mA f.s.d.	20 A	30 V	50	Approx. square law†	d.c. → ~1K	True r.m.s.
Rectifier	0.5 → 2.5	10 μA f.s.d.	0.015 A	2 A	1000 +	Approx. linear	~10 → 10K	Mean sensing
Thermocouple	0.1 → 2.5	2 mA f.s.d.	10 A	1 V	100 +	Square law	d.c. → 100K	True r.m.s.
Electrostatic	0.5	—	—	50 V	→ ∞	Square law	d.c. → 1M	True r.m.s.
d.c. direct coupled	1 → 5	—	—	1 mV	1 MΩ	Approximately linear	d.c. → 100K	may be used as narrow band amp. if output available
d.c. chopper	2.5	—	—	3 μV	30 MΩ	Approximately linear	d.c.	
a.c. mean sensing	2.5 → 5	—	—	100 μV	6 kΩ → 20 MΩ	Approximately linear	~1 → 10 ... 1M	
a.c. peak sensing	2.5	—	—	0.5 V	10 MΩ	Approximately linear	~10 → 100M	higher freq. by sampling
a.c. true r.m.s. sensing	2.5 → 5	—	▮	1 mV	30 kΩ → 10 MΩ	Approximately linear	~1 → 10M	

(ELECTRONIC — rows from d.c. direct coupled downwards)

*The magnitudes given are only an indication of general trends and *must not* be taken as applying to every instrument of a particular type.
†This applies if the instrument is used as either a voltmeter or an ammeter; when used as a wattmeter the scale is approximately linear (see reference 12).

Features

(a)   May be used for constant or slowly varying fields.
(b)   The probe or active area may be small, giving a field strength measurement over a small area.
(c)   Makes a measurement of flux density, that is, amount of flux through a known area.
(d)   The semiconductor is temperature sensitive and temperature compensation must be included in the sensing circuit.
(e)   The semiconductor is fragile and therefore easily damaged, it is also pressure sensitive.

Applications

(a)   It may be used to investigate leakage fields in air around electromagnetic apparatus.
(b)   It may be used to investigate alternating fields if the voltage output from the semiconductor is rectified before applying it to the display instrument; alternatively the flux waveform could be displayed on an oscilloscope.

Another instrument that may incorporate the Hall effect is the 'Poynting Vector Wattmeter' used for measuring the power loss density at the surface of a magnetic material, and therefore has application in the study of the iron losses of transformers and rotating machines[15,16].

1.13 INSTRUMENT COMPARISONS

Table 1.1 is a comparison of the characteristics of the commonly used analog (pointer) instruments. It should be apparent from this table that none of the instruments so far described are usable for measuring very low frequency quantities; also, the 'electronic' instruments have a much higher input impedance than the electromechanical instruments. It must however be emphasised that the magnitudes given in the table are only an indication of general trends and must not be considered as applying to every instrument of a particular type.

REFERENCES

1   C. T. Baldwin. *Fundamentals of Electrical Measurement,* Harrap, London (1961)
2   E. W. Golding and F. C. Widdis. *Electrical Measurements and Measuring Instruments,* Pitman, London (1963)
3   E. Franks. *Electrical Measurement Analysis,* McGraw-Hill, New York (1959)
4   M. B. Stout. *Basic Electrical Measurements,* Prentice Hall, Englewood Cliffs, N. J. (1960)

5   A. H. M. Arnold. 'Audio Frequency Power Measurements by Dynamometer Wattmeters', *Proc. I.E.E.* **102**, 192–203 (1955)

6   F. J. Wilkins, T. A. Deacon and R. S. Becker. 'Multi-junction Thermal converter, an Accurate d.c./a.c. Transfer Instrument', *Proc. I.E.E.* **112**, No 4 (April 1965)

7   *Which A.C. Voltmeter?; Which d.c. Voltmeter?*, Hewlett Packard application notes AN60 and AN69 (1965)

8   Sol.D. Prensky. *Electronic Instrumentation,* Prentice Hall, Englewood Cliffs, N. J. (1963)

9   D. R. Duckworth. *A.F. Power Measurement,* Marconi Instrumentation (Sept. 1968)

10   J. L. Ferns. *Meter Engineering,* Pitman, London (1932)

11   A. E. Knowlton. *Electric Power Metering,* McGraw-Hill, New York (1934)

12   H. Buckingham and E. M. Price. *Principles of Electrical Measurement,* English Universities Press, London (1955)

13   A. R. Blanchard and B. D. Lazenby. 'Maxiscale, Design for the Seventies', *Kent Technical Review* (Oct 1970)

14   *Microwave Power Measurement.* Hewlett Packard Application note A.N. 64 (1968)

15   F. D. Cocks and F. H. N. Nagg. 'Measurement of Iron Losses in a Single Lamination at High Flux Densities with a Hall Effect Wattmeter. Magnetic materials and their application.' *I.E.E. Conference Publication No. 33* (Sept. 1967)

16   H. Hollitcher. 'Hall Effect Devices and Non-destructive Testing', *A.I. Ch. E. Symposium.* Philadelphia (April 1969)

17   M. Sucher and J. Fox (Eds). *Handbook of Microwave Measurements,* Wiley, New York (1963)

18   B. M. Oliver and J. M. Cage. *Electronic Measurements and Instrumentation,* McGraw-Hill, New York (1971)

19   British Standard 89: Part 1:1970, *Specification for Direct Acting Electrical Indicating Instruments and their Accessories,* British Standards Institution, London (1970)

2

Analog (Graphical) Instruments

The majority of the instruments described in Chapter 1, while giving a continuous indication of the measurand, have required the presence of an operator to observe variations in reading magnitude. This limitation is overcome in some of the graphical recording instruments, in particular those designed to record permanently variations in the level of a quantity, and with the ever-increasing emphasis on automation, continuously recording instruments are finding many applications, temperature recorders being but one example.

Other forms of graphical recorder produce a temporary record, possibly only a display of the instantaneous values of an input signal, and these instruments which facilitate the detailed study of waveforms and analysis of circuit performance are essential for any laboratory.

2.1 MOVING COIL RECORDERS

These are an adaptation of the moving coil instrument, the scale of the instrument being modified so that a chart may be driven at constant speed by an electric or clockwork motor, under the modified pointer. The type of chart used in a moving coil recorder will depend on the form of movement. Some instruments use a chart whose horizontal axis consists of a series of circular arcs, allowing direct use of the simple moving coil movement (figure 2.1). Some instruments employ a linkage system in the movement which results in a horizontal movement of approximately straight lines[16]. The moving coil recorder is a comparatively inexpensive instrument having a narrow bandwidth (d.c. → 10 Hz) and a maximum sensitivity of about 4 mV/cm, or for an instrument with a 10 cm chart width a f.s.d. of 40 mV without amplification. Multichannel instruments are manufactured in which a number of movements operate side by side over a multitrack chart[1,2]. To obtain the permanent trace on the chart one of the following may be used.

Pen

This requires that the modified pointer shall support an ink reservoir and pen, or

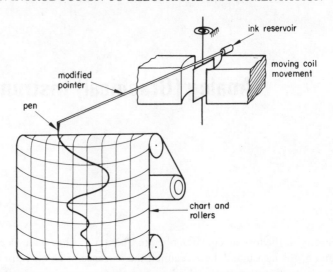

Figure 2.1. Components of a moving coil pen recorder

contain a capillary connection between the pen and a reservoir (figure 2.1). In general red ink is used but other colours are available and in an instrumentation display a colour code may be adopted. The frequency limit of recorders incorporating this method of writing is generally of the order of a few hertz, but by using a paper with a waxed surface and a special pen, sophisticated recorders are marketed that have frequency responses up to 40 Hz.

Chopper Bar

If a chart made from pressure sensitive paper is used a simple recording process is possible where the pointer is V shaped and passed under a chopper bar which depresses it into the paper once per second (or at some other suitable time interval) thus making a series of marks on the special paper. Such a system is not a truly continuous record but is suitable for recording some slowly varying quantities, for example those having a variation of 1 cycle per hour, and has the advantage of a straight line horizontal scale without the use of complex linkages.

Heated Stylus

This form of recorder requires thermosensitive paper and a pointer modified to carry a heated stylus. It is slightly more complex than either of the above in that to obtain uniform density of trace over a d.c. to 50 Hz frequency range stylus temperature control circuits must be incorporated.

Electrostatic Stylus

Using an electrosensitive recording paper and a stylus producing a high voltage discharge, another form of unprocessed permanent trace is obtained. Such a writing method has been incorporated into one commercially available recorder, which has a 5 cm wide chart; nine voltage ranges from 100 mV/cm to 50 V/cm; eight chart speeds, from 30 cm/s to 1 cm/min and a frequency response from d.c. to 60 Hz at maximum amplitude ± 1 dB.

2.2 POTENTIOMETRIC RECORDERS

The main limitations of the moving coil recorder are a moderately low input impedance and a limited sensitivity. It was shown in chapter 1 that the simple method of overcoming the input impedance problem is to position an amplifier between the input terminals and the display or indicating instrument. This technique, while producing a high input impedance and improved sensitivity, results however in an instrument of limited accuracy. This latter problem can be overcome by comparing the input signal with a reference voltage using a potentiometer circuit (figure 2.2) (see page 77). The error signal, or difference

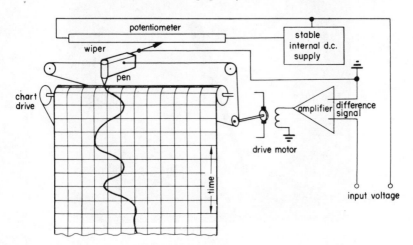

Figure 2.2. Potentiometric pen recorder

between the input level and the potentiometer voltage, is amplified and used to energise the field coils of a d.c. motor mechanically connected to the wiper on the potentiometer, which is driven in the appropriate direction to reduce the magnitude of the error signal and attain balance. These techniques result in graphical recorders with a high input impedance (tending to infinity at balance), a sensitivity maximum of 4 mV/m, an intrinsic error of less than ± 0.25 per cent, and a bandwidth extending from d.c. to 6 or 8 Hz.

Probably the commonest use of potentiometric pen recorders is for the recording and control of process temperatures. In this type of application the output from a thermocouple (see page 267) forms the input of the recorder, the scale of which would be calibrated in a suitable range of temperature, for example 0–100°C. Now the output from a thermocouple is nonlinear and thus if it desired to record on a linear scale, linearisation of the thermocouple output must be performed within the recorder circuits (see page 277). The chart for such a temperature recorder may be rectangular or circular. This latter form is the one which commonly incorporates control switches (figure 2.3) one revolution of the chart recording level variations over 24 hours or seven days.

Figure 2.3. Potentiometric pen recorder with circular display (Courtesy Honeywell Ltd)

Rectangular strip chart recorders do not usually incorporate control features but many models are available in which one recorder may be use to log several inputs. This feature may be performed either by having several (4 max.) pens which overlap each other and record the inputs simultaneously, or by replacing the pen by a print wheel geared to a selector switch so that when a particular input is connected to the balancing circuit, a point plus its identifying character is printed on the chart. This form of multipoint recorder may have as many as 24 inputs, displaying traces in as many as six colours.

The chart drive for most potentiometric recorders is derived from a motor synchronised to the power line frequency, a selection of speeds being obtained by use of a gear train.

2.3 EVENT RECORDERS

These are a simplified form of pen recorder, their operation being two state, that is the presence of an input signal causes a pen to register the beginning and its removal at the end of an event, either by the pen being in contact with the chart only for the duration of the event or by a stepped displacement of the trace for

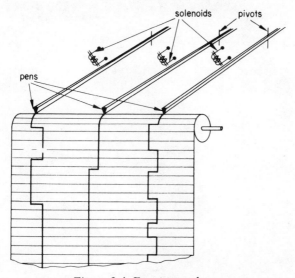

Figure 2.4. Event recorder

the duration of the event. Figure 2.4. illustrates such a trace. Such recorders are multichannel, having as many as 50 channels, and their application is in the recording of manufacturing process sequences.

2.4 X–Y PLOTTERS

The preceding types of recorder have all had time as one of their axes; however, a number of applications exist where it is desirable that both axes should be capable of representing some function other than time. To devise such a recorder it is necessary that the pen is free to move simultaneously in two directions, at right angles to each other. (The other possibility, that of moving the chart, is sometimes used). By using two potentiometer systems of the type used in the potentiometric recorder, and a mechanical transport system of the form in figure 2.5, an $X–Y$ plotter may be produced with a sensitivity on both axes of 100 μV/cm, a slewing speed of 150 cm/s, and a frequency response (for full scale deflection) on either axis of 6 Hz, the plotting area being typically 25 × 18 cm.

The writing speed of one axis when recording a sine wave signal will be π × frequency × peak to peak deflection and will have a maximum value at the upper frequency limit of the recorder. The slewing speed is the maximum writing

Figure 2.5. $X-Y$ recorder on which has been placed an inverted carriage so as to show the wiper assembly and the resistive potentiometer track (Courtesy Bryan Southern Instruments Ltd.)

speed on a 45° line, that is both servo systems driving the pen, and should equal √2 times the maximum writing speed on each axis. Such a recorder would normally include voltage range switches on both axes, the facility to convert the X axis to a time dependent scale, and internal zero offset voltages, all of which increase the versatility of the $X-Y$ plotter. The accuracy of $X-Y$ plotters is similar to that of most potentiometric recorders being approx. ± 0.3 per cent of the fiducial value.

Application

$X-Y$ plotters are used in any situation where it is desirable to save time and effort by arranging for the measuring system to produce its own graphical records. Typical applications are the plotting of stress—strain curves, semi-conductor device characteristic curves, hysteresis curves, vibration amplitude against swept frequency, plotting the output from electronic calculators or computers, etc.

2.5 ULTRAVIOLET (u.v.) RECORDERS

These are developments from the Duddel oscillograph[14], the main modifications being (a) the use of u.v. instead of white light and the recording paper that is used, (b) the housing of the galvanometers in a common magnet block instead of in individual magnets, and (c) a consequent reduction in size and weight.

Principle of operation

Ultraviolet light directed on to mirrors attached to moving-coil galvanometers is reflected via a lens and mirror system on to the special u.v. light sensitive paper, which is driven past the moving light spot thus forming a trace of current variations with time. In most u.v. recorders it is possible to select a paper speed from the 12 or so available; additionally, in some u.v. recorders the speed of the paper may be controlled by an externally applied voltage. An outline diagram of

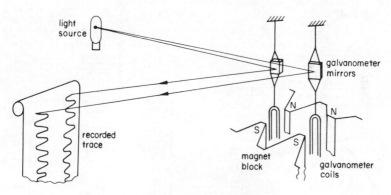

Figure 2.6. Outline of u.v. recorder operation

an instrument is given in figure 2.6. In addition to the traces of the input voltages, the following may be added to the record:

(a)   *Grid lines.* Lines along the length of the paper, obtained by shining the u.v. light through a 'comb' on to the paper.
(b)   *Timing lines.* These may cross the full width of the paper, or merely project from the edge of the paper and are derived from a vapour tube

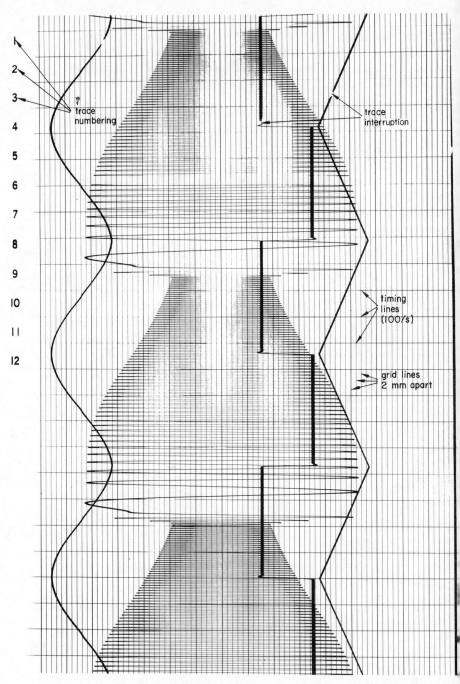

Figure 2.7. Trace from Bell & Howell 5–137 u.v. recorder fitted with 7–320 fluid damped galvanometers. Kodak paper was used at a speed of 100 cm/s

energised from either an internal source of known frequency or from
an external source.

(c)   *Trace identification.* Since some u.v. recorders are manufactured with
as many as 25 channels, each of which can produce a 100 mm peak to
peak trace on a twelve inch width of paper, considerable overlapping
of traces may result. To simplify the identification process, each trace
is momentarily interrupted in turn; coinciding with this interruption a
numeral is printed at the side of the record by directing u.v. light
through cutouts of the numerals. Figure 2.7 shows a typical record from
such an instrument.

The u.v. light sensitive paper may be processed in one of several ways:

(a)   It may be photodeveloped, that is subjected to additional u.v. light,
the traces appearing and giving a trace in 10—30 s (depending on the
level of the u.v. light). Such a record will remain in a usable state for
a considerable time providing it is not subject to excessive u.v. radiation.
Improved papers that became available during 1972 have better contrast,
and are less susceptible to the effects of additional u.v. radiation.

(b)   It may be permanised, this being a developing process similar to the
normal photographic one, that is a chemical developing and fixing process.
Some papers require photodeveloping before permanising to obtain the
best results, whilst others give the best results if additional exposure to
u.v. light is avoided; in consequence, the maker's instructions for a
particular paper must not be considered to apply to all papers.

(c)   If a record is obtained with a good contrast by photodeveloping, it is
possible to obtain short lengths of permanent record by some of the
photocopying processes. If this is done, the record will quickly
deteriorate (due to the additional exposure to u.v. light), however once
one photocopy has been obtained any number may be made by
photocopying the photocopy.

Recorder galvanometers

These are often described as pencil galvanometers (many makes being about
75 mm long and 3 mm to 6 mm in diameter), and are fitted side by side into
the recorder magnet block. As indicated above they are moving coil galvanometers, having a small mass and inertia, a restoring or control torque provided
by the ribbon suspension, and a damping torque provided either by electromagnetic means, that is depending on the resistance of the external circuit, or
by internal fluid damping[3,4,5]. Figure 2.8 shows the main features of a typical
recorder galvanometer in which a moving coil system suspended by a torsion
strip is mounted in a cylindrical frame, housed in a cylindrical container. The
surface-aluminised mirror attached to the moving coil system, coincides with a
lens to focus the light projected on to it. The aluminium wire coil is wound as a
loop around two formers, the number of turns determining the galvanometer

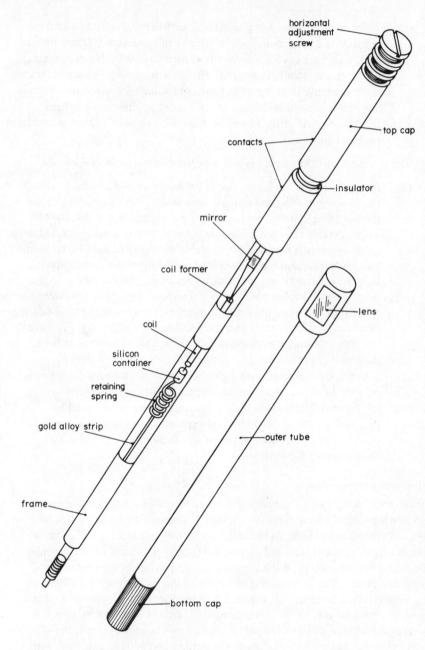

Figure 2.8. Recorder 'pencil' galvanometer (S.E. Laboratories)

sensitivity. One end of the coil is taken to the frame and the other to the top
insulated section of the container. It should be noted that although one end
of the galvanometer coil is connected to its frame, galvanometers in a common
magnet block are electrically isolated from each other.

A complete understanding of the performance of recorder galvanometers is
essential in order that true records of waveforms are obtained when using a
u.v. recorder. It is indicated above that the u.v. recorder galvanometer is a mass–
spring system and, as such, will have a resonant frequency, that is at some
frequency dependent on its moment of inertia, suspension stiffness, and mass, it
will behave as a vibration galvanometer (see page 12). This characteristic in a
device to obtain records of waveform is extremely undesirable and must be
removed by suitably damping the movement of the galvanometer.

Damping

Two methods of damping are common in connection with u.v. recorder
galvanometers, namely (a) fluid, and (b) electromagnetic damping. The amount of
damping used is also of vital importance, for if the galvanometer is overdamped
its deflection amplitude will decrease as frequency is increased, while an underdamped galvanometer will produce excessive deflections for frequencies near its
natural or resonant frequency. Thus a compromise must be made, and curves of
deflection amplitude against frequency ratio (figure 2.9) show that if a fraction
of critical damping $(\eta) = 0.64$ is used, the deflection amplitude is within ± 5 per
cent of the ideal signal amplitude for frequencies up to 60 per cent of galvanometer

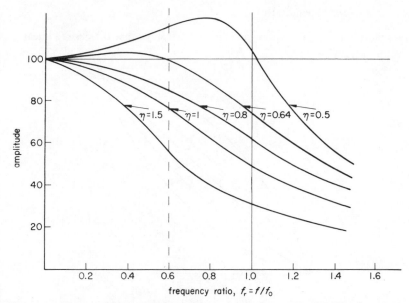

Figure 2.9. Trace amplitude/frequency ratio characteristics for a u.v. recorder
galvanometer.

resonant frequency for a sinusoidal input signal. The expression used to obtain the curves in figure 2.9 is:

$$\frac{A_a}{A_i} = \frac{1}{[(1 - f_r^2)^2 + (2f_r\eta)^2]^{\frac{1}{2}}} \tag{2.1}$$

(For derivation see appendix IIc)

where
A_a = the actual amplitude of the trace deflection,
A_i = the ideal (or d.c.) amplitude of the trace deflection,
f_r = the ratio of signal frequency (f) to the galvanometer resonant frequency, (f_0)
η = the fraction of critical damping used on the galvanometer.

In addition to the amplitude characteristics described above, some phase shift, equivalent to a time delay, must inevitably occur between the input of the electrical signal and the movement of the light spot on the photographic paper. This phase shift is unavoidable — due to the inertia and damping of the galvanometer a finite time will exist between the variation of the input signal level, and the movement of the light spot on the photographic paper — but it is of negligible importance provided it can be made directly proportional to frequency, that is providing that all frequencies of input signal are delayed by the same time interval. The relationship between phase shift (ϕ) and frequency ratio (f_r) may be shown to be:

$$\phi = \tan^{-1} \frac{2f_r\eta}{1 - f_r^2} \tag{2.2}$$

(See appendix IIc)

Curves plotted using this expression (figure 2.10) show that the desired linear

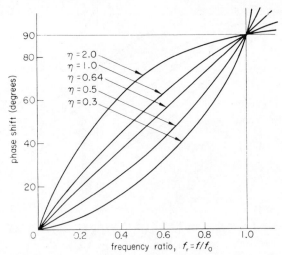

Figure 2.10. Phase shift/frequency ratio characteristics for a u.v. recorder galvanometer

relationship is approximately obtained with $\eta = 0.64$, and this, fortunately, is the fraction of critical damping which gives the optimum amplitude performance for a recorder galvanometer.

To illustrate the effect of frequency and η value on the phase shift in a recording, first consider a 50 Hz signal applied to a galvanometer that has a resonant frequency of 500 Hz and is provided with fractional damping of 0.64. Then

$$\phi_1 = \tan^{-1} \frac{2 \times 0.64 \times 0.1}{1 - 0.01} \quad \text{(from 2.2)}$$

$$= 7° \ 22'$$

or

$$\phi_1 = \frac{7.366 \times 1000}{360 \times 50} \ \text{ms}$$

$$= 0.41 \ \text{ms or } 410 \ \mu\text{s}$$

If now a 250 Hz signal were applied to the same galvanometer then the phase shift becomes

$$\phi_2 = \tan^{-1} \frac{1.28 \times 0.5}{1 - 0.25} = 40° \ 28'$$

or

$$\phi_2 = \frac{40.467 \times 1000}{360 \times 250}$$

$$= 0.45 \ \text{ms or } 450 \ \mu\text{s}$$

Thus a difference or phase error in a recording of these two frequencies with the suggested galvanometer would be 40 μs, and if a paper speed of 250 cm/s were used in obtaining the record, a displacement of

$$\frac{10 \times 2500}{1\ 000\ 000} = 0.1 \ \text{mm or 0.2 per cent of the length of the waveform}$$

as recorded for the 50 Hz wave would result. Such an error is of the same order of magnitude as the resolution of measurement using a chart reader with vernier scales and in most cases may be neglected, although it must be included in an error analysis of the measurement method (see page 214).

Table 2.1 illustrates some important aspects of phase and amplitude errors in u.v. records.

From the above table it is clear that the 0.64 value for η is more important from consideration of amplitude error than from a phase difference viewpoint. However one fact of considerable importance in relation to phase differences that does emerge is that if different galvanometers are used to measure the same or different frequencies on the same record considerable phase error may result. For example, if the 50 Hz signal were recorded using a galvanometer with a resonant frequency of 500 Hz, and the 250 Hz signal recorded using a 1000 Hz galvanometer (both with 0.64 damping) then a time difference of 200 μs or 1 per

Table 2.1

signal freq. (Hz)	galvanometer resonant frequency (Hz)	fraction of critical damping	phase angle (θ)	time lag (μs)	time diff. (μs)	amplitude difference from d.c.
50	500	0.5	$5°\ 46'$	320	⎫ '54	+0.5%
250	500	0.5	$33°\ 41'$	374	⎭	+10.9%
50	500	0.64	$7°\ 22'$	410	⎫ 40	+0.2%
250	500	0.64	$40°\ 28'$	450	⎭	+1.4%
50	500	0.8	$9°\ 11'$	510	⎫ 11	−0.3%
250	500	0.8	$46°\ 51'$	521	⎭	−8.8%
50	1000	0.64	$3°\ 40'$	204	⎫ 5	+0.04%
250	1000	0.64	$18°\ 51'$	209	⎭	+0.95%

cent of the 50 Hz waveform length would be obtained. This is an unfortunate shortcoming as it is often desirable that galvanometers of different sensitivities and hence different frequency ranges be used in the compiling of a record. The most satisfactory method of overcoming this problem is to use only one type of galvanometer and connect adjustable gain amplifiers between the small amplitude signals and the galvanometers.

Transient response

The above description has been concerned with the performance of the galvanometers when supplied with a signal which has sinusoidal waveform. However, a number of very suitable applications for using the u.v. recorder are transient in nature. It is therefore important to be aware of the limitations of galvanometer performance under transient conditions. The simplest, yet most severe, form of transient is a step function of voltage. Applying such a transient to a u.v. recorder galvanometer will cause a deflection of the spot light to overshoot the steady state deflection x, by an amount y and the magnitude of this overshoot will depend on the amount of galvanometer damping present as illustrated by table 2.2, in which the percentage overshoot = y/x . 100 per cent, and η = the fraction of the critical damping applied to the galvanometer.

Table 2.2

% overshoot	η	% overshoot	η
0.1	0.911	8.4	0.619
0.5	0.861	9.6	0.597
1.2	0.815	10.8	0.577
2.4	0.765	13.5	0.537
4.0	0.716	16.5	0.497
5.2	0.685	19.5	0.461
6.4	0.658	22.5	0.429
7.3	0.64	25.0	0.404

It is apparent from table 2.2 that to obtain a record without overshoot requires the galvanometers to be used with critical damping, but such a mode of operation is usually unacceptable because of the time delay between the change in signal level and the movement of the light spot on the record. Figure 2.11 shows the

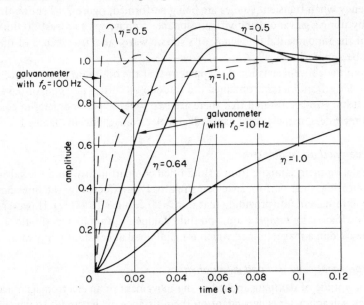

Figure 2.11. Galvanometer response to a step function

displacement–time characteristics for a galvanometer with a resonant frequency of 10 Hz and various values of damping when supplied with a step voltage. Also shown in this diagram are the characteristics for a galvanometer with a resonant frequency of 100 Hz with fractional damping values of 0.5 and 1.0. It should be observed that the time to steady deflection for this latter galvanometer is one tenth that for the 10 Hz galvanometer, indicating that for transient studies, galvanometers with high resonance frequencies must be used. Again it becomes

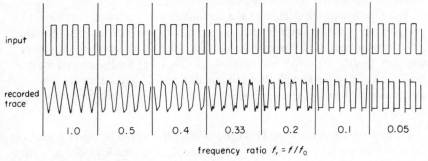

Figure 2.12. The effect of f_r on transient distortion in a recorded trace (Courtesy of Bell and Howell)

apparent that 0.64 of critical damping results in an optimum of performance, being a compromise between excessive overshoot (with its associated oscillations) and no overshoot but excessive delay in attaining correct deflection. As a further illustration of the importance of using a galvanometer with a high resonance frequency when transient studies are being performed, figure 2.12 shows that only by using a galvanometer with a resonance frequency of at least 20 times that of the fundamental frequency of a square wave can a true record of the signal be approached.

From the above discussion it is apparent that the optimum performance from a recorder galvanometer is obtained if a fraction of critical damping of 0.64 is used. It has also been stated that there are two types of galvanometer in common use in recorders, namely fluid damped, and electromagnetically damped.

Fluid damped galvanometers

These derive their damping from a fluid filled compartment within the galvanometer casing, their performance being relatively unaffected by the impedance of the external circuit (providing that it is $> 20 \ \Omega$ but $< 2000 \ \Omega$). However, they may be affected by temperature, and fluid damped galvanometers should therefore be used in a magnet block which is maintained at constant temperature.

Electromagnetically damped galvanometer

The magnitude of damping applied to the movement of an electromagnetically damped galvanometer is dependant on the resistance of the circuit to which it is connected. For example, a galvanometer connected to a low resistance circuit will be heavily damped whilst one connected to a high resistance circuit will be lightly damped. This is best understood by considering a galvanometer not connected to a circuit; if the galvanometer movement is mechanically twisted through an angle, and then released, it will swing back to the zero position, generating a voltage at its terminals. If the terminals are connected to a resistance, a current will flow causing a force in the galvanometer opposing the change in position of the galvanometer coil, and the smaller the resistance, the larger the current will be, and so too the greater the damping force.

Since the amount of damping has a profound effect on the galvanometer performance, the magnitude of resistance that the electromagnetically damped galvanometer 'sees' is of importance. Manufacturers of recorder galvanometers therefore specify the magnitude of resistance (R_D) that a galvanometer must 'see' in order to operate with the optimum of 0.64 of critical damping. There are three basic forms of circuit in which a galvanometer is likely to be operated, namely:

(a) *Low impedance source.* This condition is represented by figure 2.13a, where the signal source has a resistance R_s lower than the required R_D value. It is therefore necessary to insert a resistance R_a in series with the source so that the resistance the galvanometer 'sees' is

$$R_D = R_s + R_a$$

that is $\qquad R_a = R_D - R_s \ \Omega$

(b) *High impedance source.* To obtain the correct R_L value for this condition, (figure 2.12b) a resistance R_b must be connected across the source such that:

$$R_D = \frac{R_b \cdot R_s}{R_b + R_s} \quad \text{or} \quad R_b = \frac{R_D \cdot R_s}{R_s - R_D} \ \Omega$$

(c) *A source requiring a matched load.* This is a more general condition than either of the above two, and arises when, for example, a galvanometer is supplied

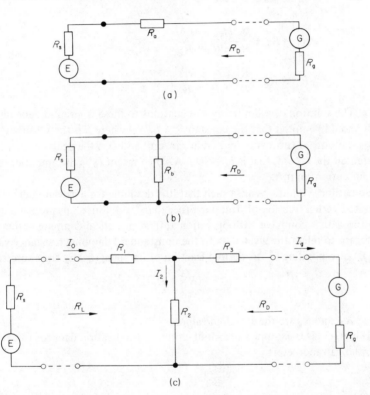

(a)

(b)

(c)

Figure 2.13. Galvanometer damping circuits. (a) Low impedance source (b) high impedance source (c) attenuator to match both source and galvanometer impedances

via an amplifier. The magnitudes of the resistors R_1, R_2, and R_3 in figure 2.13c are determined by evaluation of the equations:

$$R_1 = R_L - \frac{R_2 \ (k-1)}{k} \qquad (2.3)$$

$$R_2 = \frac{k(R_L + R_s)(R_D + R_g)}{k^2(R_L + R_s) - (R_D + R_g)} \qquad (2.4)$$

and

$$R_3 = R_2(k-1) - R_g \qquad (2.5)$$

The derivation of these equations is obtained by the solution of the network equations, that is

$$I_0 = I_2 + I_g; \quad I_g = \frac{R_2}{R_3 + R_g}; \quad I_0 = I_g\frac{(1 + R_3 + R_g)}{(R_2)}$$

or

$$\frac{I_0}{I_g} = 1 + \frac{R_3 + R_g}{R_2} = k \qquad (2.6)$$

$$R_L = R_1 + \frac{R_2(R_3 + R_g)}{R_2 + R_3 + R_g} \qquad (2.7)$$

$$R_D = R_3 + \frac{R_2(R_1 + R_s)}{R_1 + R_2 + R_s} \qquad (2.8)$$

Note. The solution of a similar set of equations to these is given in appendix III.

It should be noted that the constant $k = I_0/SD$ where S is the sensitivity of the galvanometer (normally quoted in mA/cm) and D is the peak deflection desired on the record, that is $I_g = SD$ mA. The current I_0 is the amplifier or source output current in mA.

Sometimes conditions arise such that it is desirable for a galvanometer to be operated with a fraction of damping other than 0.64. Since galvanometer specification data is compiled with operation at 0.64 of critical damping in mind it is necessary to relate the given data to the nonstandard damping resistance value (R_d), corresponding to the desired fraction of damping (η') by the expression

$$R_d = \frac{0.64(R_D + R_g)}{\eta'} - R_g \quad \Omega \qquad (2.9)$$

where R_D and R_g are the specified values.

Table 2.3 gives examples of manufacturer's specification data for two u.v. recorder galvanometers:

Table 2.3

Type No.	External damping resist. $R_D(\Omega)$	Natural freq. f_0(Hz)	Flat (5%) freq. range f(Hz)	Terminal resist. (± 10%) $R_g(\Omega)$	d.c. sensitivity S (mA/cm)	Maximum safe current (mA)	Damping
7–349	350	10	0–6	130	0.00065	15	electromagnetic
7–362	20–2000	4150	0–2500	69	4.72	100	fluid

(Reproduced by kind permission of Bell & Howell Ltd)

The recommended maximum deflection for either galvanometer is ± 5 cm, so the peak galvanometer current (I_g) for this deflection using a type 7–349 should be ± 3.24 μA, and for the 7–362, ± 23.6 mA, which indicates the range of galvanometer sensitivities available. The upper limit of flat frequency response of u.v. recorder galvanometers is approximately 8 kHz, and at a paper speed of 10 m/s this would produce a trace having a distance between peaks of 1.25 mm which is insufficient for a sensible study of the waveform. These high frequency galvanometers are however vital in transient studies. The limit of frequency to which a u.v. recorder can be used will depend on the magnitude of the displayed trace, and the paper speed, but as a guide a trace with a peak to peak deflection of 5 cm is recognisable as a sine wave providing the length of one cycle is greater than 1 cm.

Galvanometer selection factors

In selecting a galvanometer it is important that one is chosen so that the following are complied with.

(a)   a flat frequency range in excess of the frequency of the highest harmonic of importance in the a.c. signal, (a low frequency galvanometer may be used as a low pass filter to remove unwanted 'noise' from a signal). Alternatively if transients are being studied, the galvanometer's resonant frequency should be at least 20 times that of the 'fundamental' frequency of the transient, thus demonstrating the necessity for the manufacture of galvanometers with high resonant frequencies, for example a galvanometer with an f_0 of at least 1000 Hz should be used when recording the effects of a step change in the load on a 50 Hz supply.

(b)   a sufficiency of sensitivity so that a satisfactory magnitude of deflection on the record may be obtained. It being appreciated that the larger the trace the smaller the error due to the uncertainty in measuring deflections on the trace.

The final selection of a galvanometer will invariably be a compromise between the two requirements given above, and the limitations of the circuit to which the galvanometer is to be connected. From the specification data tabulated above it is apparent that the sensitivity of a galvanometer is inversely proportional to the flat frequency response, and a galvanometer with a sufficiency of sensitivity and frequency range may be unobtainable. This unfortunate fact can only be overcome by the insertion of an amplifier between the signal source and the galvanometer. Amplifiers specifically built for this purpose are available, some with a continuously variable gain, and others with calibrated steps of gain. In both cases the amplifier output usually contains current limiting and galvanometer matching components.

Calibration

Having selected a suitable galvanometer and matching circuit, the most satisfactory method of calibrating the obtained trace in terms of the absolute magnitude of signal to be measured is to replace it by a signal of known magnitude and frequency. If several levels of known or calibrating signal are used to obtain short lengths of trace, a mean signal to deflection constant of calibration may be derived. This procedure must be followed for each galvanometer and its matching circuit, for each application of the recorder.

Applications

It has already been indicated that the range of frequencies over which the u.v. recorder can generally be used is for d.c. and a.c. signals having a fundamental frequency of up to 400 or 500 Hz depending on the paper speeds available on the recorder being used. The recording of higher frequencies is possible on the few recorders with very high (10 m/s) paper speed. The application of the recorder is in any situation where a trace of a waveform is required, and resolves to the problem of scaling the signal to be measured to a suitable magnitude, whilst ensuring that the matching requirements of a particular galvanometer are met.

The commonest method of scaling voltages is to use a resistance divider (see page 153). Current waveforms are obtained by recording the voltage drop across a known resistance. Having recorded traces of voltage and/or current waveform it is possible to determine their frequency and any phase differences between traces (using the known time scale projected on to the record by the time markers). Remembering that the resolution of reading is of the order of 0.3–0.5 mm and corrections for unequal galvanometer phase shift must be included.

Typical applications of the u.v. recorder are in recording regulation transients of generators, the performance of control systems, the outputs of transducers, and the magnitudes of low frequency signals which cannot be measured by pointer instruments.

2.6 THE CATHODE RAY OSCILLOSCOPE (C.R.O.)

These are familiar pieces of measuring equipment, as no TV programme with a science content is complete without the display of instantaneous voltage values on an oscilloscope. It is an indispensable instrument in any laboratory where voltage waveforms have to be observed and it is manufactured in variations of complexity from a simple basic instrument to the sophisticated programmable instrument with digital readout, figure 2.14. Since whole books (for example

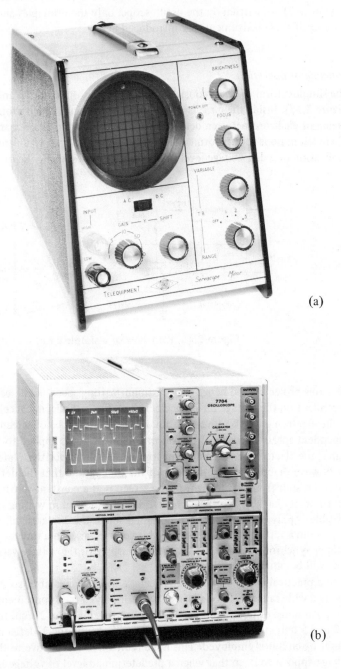

Figure 2.14(a). Simple and (b) complex cathode ray oscilloscopes (Courtesy Telequipment Ltd and Tektronix (U.K.) Ltd.)

Reference 7) are written on the oscilloscope, only the principles and main features of these instruments are introduced here.

Principle of operation

The simplest form of c.r.o. consists of a cathode ray tube (c.r.t.) and a timebase (figure 2.15). Inside the c.r.t. is situated an electron gun which projects a fine stream of electrons between deflecting plates on to a phosphor coated screen, where a luminous spot is formed. The focusing of this spot is controlled by the application of a d.c. voltage level to biasing electrodes, while the quantity of

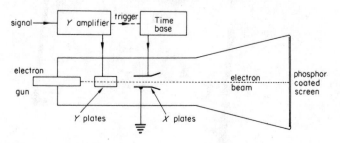

Figure 2.15. Plan view of a simple c.r.o.

electrons projected and hence the maximum writing speed of the oscilloscope is dependent on the voltage difference between the cathode and accelerator electrodes of the electron gun. To move the spot across the tube face or screen at a time dependent speed a 'saw tooth' waveform (figure 2.16) is applied to the X deflection plates, the electron beam being bent towards the more positive plate. The saw tooth wave shape is used as the basis of the timebase so that the deflection of the luminous spot from left to right of the screen is at a constant velocity, whilst the return or 'fly back' is at a speed in excess of the maximum writing speed and hence invisible. Applying a difference in potential across the Y plates will cause the spot to move in a vertical direction, and if this voltage varies in a time dependent manner synchronised with the timebase, a display of the voltage variations with time will be obtained on the screen.

In a practical oscilloscope[6,7] the timebase will be adjustable, so that signals having a wide range of frequencies may be displayed on a convenient time scale. A typical range of horizontal deflection sweeps being from 2 s/cm to 1 μs/cm, in 1, 2 and 5 unit steps. To synchronise the timebase and the Y deflecting signal, a triggering circuit is employed. This is a circuit which is sensitive to the level of voltage applied to it, so that when a predetermined level of voltage is reached, a pulse is passed from the trigger circuit to initiate one sweep of the timebase. Thus on a timescale a series of events as shown in figure 2.16 are occurring within

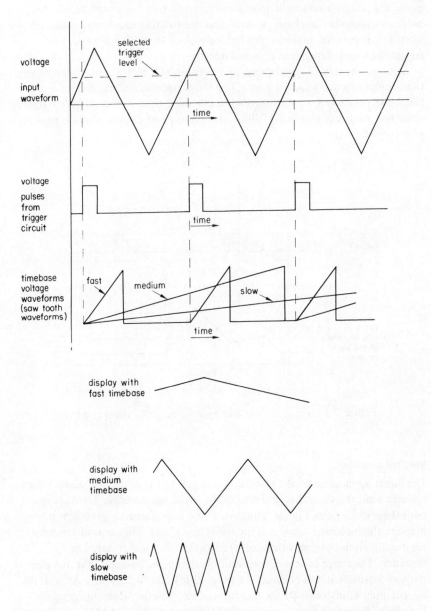

Figure 2.16. Effects of triggering and timebase selection on the c.r.o. display

the oscilloscope. The trigger circuit of an oscilloscope will be adjustable so that a particular point on either the positive or negative half cycle may be selected, and used, to trigger the timebase. As an alternative to using the display signal to internally trigger the timebase the timebase of most oscilloscopes may be triggered by an independent external signal.

Some of the more sophisticated oscilloscopes[6,7,8] incorporate sweep magnification, and sweep delay facilities, enabling the expansion and examination of particular points in a waveform which occur some time after a suitable triggering point has occurred. Figure 2.17 illustrates the type of display which is possible with these facilities.

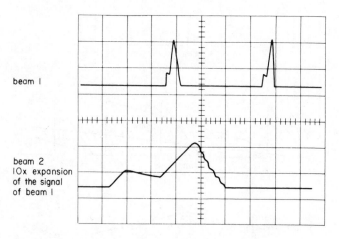

Figure 2.17. c.r.o. display using sweep magnification and delay

Vertical amplifiers

The direct application of the vertical or Y signal to the deflection plates would severely limit the versatility and sensitivity of the oscilloscope. To overcome both these difficulties a range 'attenuator' and amplifier arrangement is inserted between the incoming signal and the deflection plates. The steps of sensitivity are usually given in terms of mV or V/cm deflection, and arranged in a 1−2−5 sequence. The range of sensitivities varies between oscilloscopes but in a general purpose instrument will probably be from 10 mV/cm to 20 V/cm. As with the electric pointer instruments the inclusion of an amplifier whilst bringing the benefits of increased input and sensitivity introduces drift problems, limitations of bandwidth, and accuracy. In general the most sensitive (100 μV/cm) oscilloscope ranges have the narrowest bandwidth (for example d.c.−50 kHz) while the less sensitive ranges of some 'scopes extend into the region of hundreds of megahertz. To remove the effects on the display of d.c. drift in the circuit under investigation most Y input amplifiers have an a.c. or d.c. mode of operation switch which enables a blocking capacitor to be inserted in the input connection thus isolating the drift voltages from the input amplifier in the a.c. mode.

The above assumes that all voltages are measured with respect to earth or ground potential (see page 140). This is not always so, since it may be desirable to display the voltage waveform across a component in a line, that is both ends of the component having a potential to earth. To overcome this problem some oscilloscopes' vertical amplifiers are manufactured with two inputs so that either input with respect to earth may be displayed or the difference between the two inputs may be displayed. This form of input amplifier is known as a differential input amplifier (see page 177).

X–Y display

As an alternative to the timebase circuits which may be connected to the X plates, most oscilloscopes have an uncalibrated X amplifier while some dual beam instruments have the facility for using one of the Y amplifiers as an X amplifier. This enables the display of loops or Lissajous figures to be made (see page 73). If time varying signals of *equal magnitude* and zero phase displacement are applied simultaneously to both the X and Y *plates* of an oscilloscope the display would be one of a straight line at 45° to the horizontal. If, however, a time lag equal to 90 electrical degrees existed between these two signals of *equal magnitude* then the display would be circular. For angles other than these two values an elliptical display (figure 2.18) is obtained in which the major and minor axes of the ellipse are related to the phase angle (θ) between the X and Y voltages by the expression $\tan \theta/2 = b/a$.

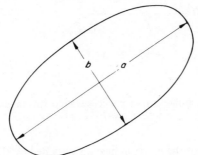

Figure 2.18. Elliptical display for phase angle measurement

Note. The X and Y displays must be of equal magnitude, and this should be checked by separately measuring the magnitudes of the displays.

'Z' modulator

The strength of the electron beam leaving the electron gun will depend on the voltage difference between its grid and the cathode. If a pulse of voltage were applied, via a blocking capacitor, to the grid it would be possible to blank out the display for the duration of the pulse. This is the purpose of the 'Z mod.' terminal available on many oscilloscopes, and enables part of a display to be either reduced or increased in luminosity.

Sampling oscilloscopes

The continuous display on an oscilloscope is, at the present time, limited to frequencies in the 50–300 MHz range, depending on the 'scope design. Above this order of frequency sampling techniques[9,10,16] must be used to obtain a satisfactory display, which may be made up from as many as 1000 dots of luminescence. The vertical deflection for each dot is obtained from progressively later points in each successive cycle of the input waveform (figure 2.19). The

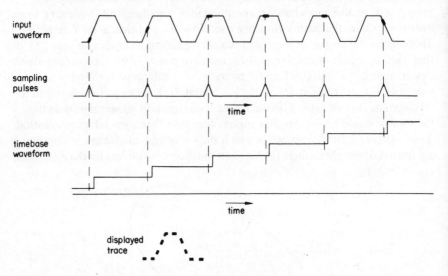

Figure 2.19. Simplified diagram of process waveforms in a sampling oscilloscope

horizontal deflection of the electron beam is obtained by the application of a staircase waveform to the X deflection plates.

Sampling techniques may not, of course, be used for the display of transient waveforms.

Screen phosphors

The screen coating material absorbs the kinetic energy of the electon beam and gives off luminous energy. Several materials[15] of different characteristics are available for this purpose, examples in common use being as shown in Table 2.4.

Table 2.4

Phosphor Type	Trace type and use
P.1	green trace of medium persistence, satisfactory for photographing, good for visual work.
P.2	bluish green trace with a long persistent yellow phosphor-escence, very suitable for studying slowly varying signals.
P.4	white trace, used in television displays.
P.7	blue-white trace, long persistence yellow-green phosphor-escence.
P.11	blue trace of short persistence, good for photographic work, mainly used in high speed oscillography.
P.31	green trace, good general purpose phosphor.

Thus the selection of a particular screen phosphor should be decided by the intended applications of the c.r.o.

Storage oscilloscope[11,12,15]

In these instruments a special phosphor may be made to retain, for viewing on command, a waveform or number of waveforms for an hour or more after the event has occurred. This feature is particularly useful when comparing a number of slowly varying signals with each other, either on the c.r.o. screen, or to produce a collective photograph. The stored waveforms can, of course, be erased by command.

Bandwidth and risetime

Of primary importance in the use of an oscilloscope, is adequate high frequency response. For 'slow' oscilloscopes this is satisfactorily specified as a bandwidth or usable frequency range and quoted as the 3 dB points — the frequency at which the displayed amplitude is 0.707 of the applied signal (see page 130) — of the vertical amplifier. For 'fast' or high frequency oscilloscopes, risetime — normally the time interval between 10 per cent and 90 per cent of the amplitude of a step function — and bandwidth should both be specified. These quantities must be compatible and a figure of merit[15] obtained from the product, risetime x bandwidth, should have a value less than 0.35 to obtain optimum transient response for a step function having a risetime greater than 5 times that of the oscilloscope, for example a 'scope with a bandwidth of 15 MHz, and a risetime of 23 ns would have a figure of merit of $23/10^9 \times 15 \times 10^6 = 0.345$ and could be used to measure, with 2 per cent accuracy, risetimes greater than 115 ns.

It has previously been stated that high sensitivity and wide bandwidth are incompatible, and so that an oscilloscope may be produced with both these desirable qualities a compromise must be made. This is attained by producing oscilloscopes with 'plug in' amplifiers thereby enabling the versatility of a single oscilloscope to be considerably increased without prohibitive cost. The ultimate limiting factor in the bandwidth of an oscilloscope is the writing speed of the spot, and for the study of very high frequency signals, sampling techniques must be adopted.

Multiple trace displays

In many instances it is necessary to compare one signal with another. To facilitate this, multiple trace displays are obtainable by:

(a)   The use of two electron guns within the same cathode ray tube. This method produces an instrument known as a *dual beam oscilloscope*, in which the electron beams of the two channels are completely independent of each other. This effect may also be produced by a single electron gun, the output from it being split into two independently controllable electron beams.

(b)   Using a single electron gun and producing a double trace by switching the Y deflection plates from one input signal to another for alternate sweeps of the screen. The eye interprets this as a continuous simultaneous display of the input signals although it is a sampled display. An oscilloscope using this technique known as *alternate mode* may only be used as a single channel instrument for recording transient phenomena.

(c)   Switching the input to the Y deflection plates from one input signal amplifier to another at a rate not governed by the duration of the sweep; that is the traces of each channel of input are built up from a series of dots as the electron beam receives a deflection derived from a particular input channel. This technique known as *chopped mode* may be used for multichannel transient investigations.

Methods (b) and (c) reduce the cost of producing a multichannel oscilloscope, it being possible to use either of these techniques to obtain a display of more than two channels.

Oscilloscope accessories

To increase the utility of an oscilloscope, various accessories are available for use with it, or built in to it.

Calibrators. Many oscilloscopes have a built in reference source of voltage, which usually takes the form of a 1 kHz, square waveform of either a single magnitude or of selectable magnitudes. This facility enables the oscilloscope timebase and amplifiers to be checked for accuracy of calibration each time the 'scope is used. Should the calibration be found to be outside the specified limits for a particular instrument, the setting up procedures stipulated in the manufacturer's handbook should be performed by authorised personnel.

Probes. This topic is dealt with in chapter 4 but briefly the necessity for their use arises from the problem of ensuring that a circuit under investigation is unaffected by the presence of the measuring instrument. If the frequency in the said circuit is high then the capacitance of the cable conveying the signal to the oscilloscope will present the circuit under test with a low impedance. To reduce the errors that this would produce, a probe (having a high impedance) must be used at the measuring end of the connecting cable.

Cameras. The best method of obtaining permanent records of oscilloscope traces for analysis is to use photography. Special cameras are available for this purpose and are of the following two types.

(a) *35 mm.* Usually a prefocused unit that is bolted over the tube face. Some cameras have a fixed lens aperture, the exposure being simply controlled by hand operation of a flap shutter. More sophisticated units incorporate aperture and exposure control. The types of 35 mm film in common use are panchromatic and special blue sensitive films, which are faster and thus more suitable for the recording of high speed transients. Both of these give a 'negative' record, that is a black trace on a transparent background, from which prints in the form of a white trace on a black background may be made. A third type of 35 mm photograph is obtainable by using recording paper, which results in a black trace on a white background. These are, however, difficult to enlarge.

(b) *Polaroid film.* This is an approach to 'instant' photography, a permanent record being obtained in 10–20 s. This process is considerably more expensive in materials than 35 mm photography, although the relative cost may be reduced by using the facility of some of these cameras to be moved relative to the screen, making it possible to place several traces (exposures) on one print. The camera is again prefocused, but normally provided with aperture and exposure controls. The record obtained on Polaroid film is a white trace on a black background, that is a print, and as such is difficult to reproduce.

Note. Time is money, and if only a few photographs are to be made, Polaroid photography is usually the most economic method.

Applications

While the oscilloscope is a very necessary and versatile instrument it must be remembered that the accuracy of direct measurements made using an oscilloscope are rarely better than ± 3 per cent on either the X or Y axes, and is therefore really a display instrument with facilities enabling the estimation of magnitudes of voltage and time.

Voltage measurement

The parameter of voltage which is most easily determined using an oscilloscope is, for a sine wave, the peak to peak value; and for a pulse its peak value. In either case the magnitude is determined using the engravings on the graticule, in conjunction with the calibrated ranges of the input amplifier.

Note. Peak to peak voltage = $2\sqrt{2}$ x r.m.s. value. (For a pure sine wave.)

Current measurement

The oscilloscope is a high input impedance instrument, and therefore cannot be directly used for the measurement of current. Currents can of course be measured as voltage drops across resistors, but care must be exercised in connecting the oscilloscope leads to a resistor for this purpose because unless a differential input amplifier is being used one side of the voltage dropping resistor will have to be at earth potential (see page 147).

Phase angle measurements

An approximate measurement of the phase angle between two voltages may be performed using the graticule of a dual trace oscilloscope to determine first, the length of a complete cycle, and then the separation of the peaks of the two voltage waveforms on the two traces. A more accurate method of phase angle measurement is to form an elliptical display, by applying one voltage to the X input and the other (of equal magnitude) to the Y input, when ϕ the phase angle between them will be equal to $2 \tan^{-1} b/a$ where a, and b, are the major and minor axes of the ellipse as in figure 2.18 (see also page 67).

Frequency measurements

The simplest method of measuring frequency with an oscilloscope, is to determine (using the graticule) the distance between two identical points in a waveform and multiply this by the timebase setting to obtain the period and hence the frequency. This method, whilst quick, is subject to at least a ± 5 per cent error. More accurate methods of frequency measurement using an oscilloscope, utilise it as a display—detector in comparison techniques, employing a stable known frequency from an external source, and the 'scope's $X-Y$ display facility.

Consider first the problem of adjusting an oscillator's output frequency to be equal to that of a standard frequency signal. Applying one of these signals to the X plates of the oscilloscope and the other to the Y plates will result in an ellipse type display which will vary in shape at a rate dependent on the difference in frequency between the two signals. As the oscillator's frequency is adjusted to be equal to the standard frequency the rotations of the ellipse will become slower and when the two frequencies are equal will cease altogether. Thus it becomes apparent that the difference between two signals which have approximately equal frequency may be determined by counting the 'rotations' of the ellipse in a known time. In this manner extremely accurate frequency comparisons may be made using an oscilloscope. For example, if an oscillator's output frequency changes with respect to a 1 kHz standard by one cycle (one rotation of the display) in 10 s then the difference in frequency between the two is a tenth of a Hz or 0.01 per cent.

For multiples of a standard frequency, Lissajous figures (figure 2.20) are used. The actual pattern of these displays will depend on the phase displacement of the signals as well as their ratio, but at a particular phase angle for each ratio of frequencies a symmetrical pattern is obtainable, and under these conditions the ratio of peaks or loops on the left hand side to those on the top edge of the Lissajous figure correspond to the frequency ratio of the signals applied to the

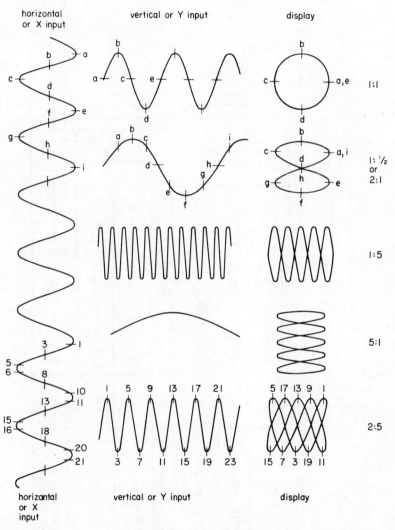

Figure 2.20. Lissajous figures and the corresponding input signals on the X and Y axes

Table 2.5. Comparison table of graphical instruments (see also table 9.1).
Note. The values are typical values and must not be taken as applying to every instrument of a particular type.

type	accuracy class	maximum sensitivity	input impedance	frequency range (Hz)	record
moving coil ink pen	1.0	4 mV/cm	100 Ω →		permanent graph of variable against time
" chopper bar	"	"			
" heater stylus	"	"			
" electrostatic stylus	"	100 mV/cm	20 kΩ		"
potentiometric ink pen	0.5	4 μV/cm	100 kΩ		"
u.v./light spot	5.0	0.65 μA/cm	130 Ω		depends on paper may need 'permanising'
"	"	4.73 mA/cm	69 Ω		
cathode ray oscilloscope	5.0	10 μV/cm	10 MΩ		visual and photograph

Frequency range (Hz) scale: d.c. 0.01 0.1 1 10 100 1K 10K 100K 1M 10M 100M 1G 10G

X and Y inputs of the oscilloscope. Large ratios of frequency are difficult to interpret, apart from the complexity of the pattern, as any slight change in one of the frequencies will cause the pattern to roll. A method of overcoming the pattern complexity problem is to use 'roulette' patterns[13].

Another method of determining frequency ratios using an oscilloscope is to supply a common frequency signal to the X and Y inputs, these voltages having a $90°$ phase shift (to give a circular display), and applying a higher frequency signal to the Z modulator input. This higher frequency signal varies the intensity of the circular display, there being one bright and one dark section, for each multiple of $X-Y$ input frequency. This method, whilst giving a display which is easy to interpret, is limited in application to exact multiples of input frequencies, with the additional requirement that the Z input frequency must be larger than the $X-Y$ input.

Risetime measurements

When pulse waveforms are being studied, a property of considerable importance is the risetime of the pulse. The obvious method of measurement of this quantity is to use the oscilloscope graticule and timebase. However if the pulse has a very fast risetime, the oscilloscope may impose one, or both, of the following limitations. First the Y amplifier bandwidth (or risetime) may be insufficient to cope with the high frequency components of the pulse risetime; and secondly, the capacitance of the cable connecting the signal to the oscilloscope may slow the risetime of the pulse and cause an incorrect display.

The first restriction can only be overcome by changing to another oscilloscope or a 'plug-in' amplifier with a higher frequency response. The second error causing condition may be removed by using a 'scope lead fitted with a frequency compensated probe, that is a probe whose input impedance is unaffected by the signal frequency (see page 135).

2.7 RECORDER COMPARISONS

Table 2.5 compares the accuracies, input impedances, and frequency ranges of the various graphic recorders. It will be observed that these instruments conveniently fill the gap in the frequency response column of the pointer instrument comparison table that is shown in Table 1.1.

REFERENCES

1   D. R. Davies and C. K. Michener. 'Graphic Recorder Writing Systems'. *Hewlett Packard Journal.* **20**, No. 2 (Oct. 1968)

2   E. Price and D. Taylor. 'Pen Recorders — a development report'. *Control and Instrumentation* (Sept. 1970)

3   M. A. Le Gette. *The Theory of Recording Galvanometers.* Published by Consolidated Electrodynamics Corporation (Bell and Howell) (1962)

4   *The Galvanometer User's Handbook.* Consolidated Electrodynamics
    Corporation (Bell and Howell, Basingstoke) (1971)
5   *Galvanometers.* S.E. Laboratories (Engineering) Ltd, Feltham
6   *Fundamentals of Selecting and Using Oscilloscopes.* Tektronix U.K. Ltd
    (1961)
7   J. Czech. *Oscilloscope Measuring Techniques.* Macmillan, London (1965)
8   *Understanding Delaying Sweep.* Service Scope No. 50. Tektronix U.K. Ltd
    (1968)
9   *Sampling Notes.* Tektronix U.K. Ltd, Beaverton (1964)
10 R. Carlson et al. *Sampling Oscillography.* Hewlett Packard. Application
    Note 36
11 Three New Instruments, 3 Kinds of Storage, *Tekscope,* **4**, No. 4, Tektronix
    Inc. (July 1972)
12 D. C. Calon. *Advances in Storage Oscilloscopes.* Electronic Industries
    (U.S.A.) (Feb. 1966)
13 *Frequency Comparisons using Roulette Patterns.* Tektronix U.K. Ltd (1960)
14 H. Buckingham and E. M. Price. *Principles of Electrical Measurements.*
    English Universities Press, London (1955)
15 *Reference Information.* Tektronix General Catalogue (1971)
16 B. M. Oliver and J. M. Cage. *Electronic measurements and Instrumentation,*
    McGraw-Hill, New York (1971)

3

Comparison Methods

The methods described in this chapter could, as an alternative, have been termed 'null methods'; the measurement process being to reduce the difference between a known and an unknown quantity to zero, that is, a null would be indicated. Such methods have inherently a greater precision than direct measurements, for example, by such methods it is possible to compare two voltages of the order of 1 V using a detector that has a resolution in terms of microvolts. The accuracy of such methods must, however, depend on the limits of error that apply to the particular 'standard' quantity.

3.1 D.C. POTENTIOMETER

Principle of operation

Consider the simple circuit in figure 3.1. The battery of voltage E drives a current I around the closed circuit ABCD. If the connection BC consists of a metre length of resistance wire having a resistivity and area such that its resistance is

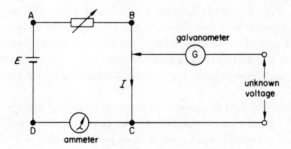

Figure 3.1. Slide wire potentiometer

20.0 Ω, and the value of I is 50 mA, the voltage drop along BC will be 1 V or 0.001 V/mm. Thus any d.c. voltage less than 1 V may be determined to three significant figures by adjusting the position of the slider to give zero deflection on the galvanometer G, and measuring the length of wire between the wiper and the end C.

The errors of the above will be largely dependent on the quality of the ammeter, and this limitation can be removed by comparing the slide wire voltage with an accurately known voltage such as that provided by a standard cell (see page 209).

A metre long slide wire is unwieldy and if two movable contacts are used, one on a shorter slide wire and the other on fixed lengths of resistance wire, a more compact instrument can be constructed. If the circuit in figure 3.1 is thus modified to that in figure 3.2, where the slide wire now has a length of 0.1 m and a resistance of 2 Ω connected in series with 15:2 Ω resistors, the total volt drop (with a current of 50 mA) will be 1.6 V.

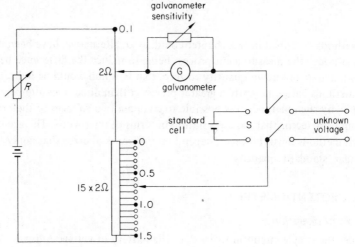

Figure 3.2. Simple d.c. potentiometer

To standardise or calibrate such a potentiometer it is necessary to connect the divider voltage, via switch S, to the standard cell. Then, with the movable contacts set at points such that the voltage appearing on the divider should be equal to the standard cell voltage, adjustment of the variable resistor R is made until the value of current in the potentiometer circuit causes these voltages to be equal, this being apparent when the galvanometer deflection (at maximum sensitivity) is zero.

The switch S can then be changed to connect the divider voltage, via the galvanometer, to the unknown and the divider contacts moved until balance is obtained — this being the basis of single range potentiometers which typically have a resolution of 1 mV and an accuracy of 0.1 per cent. They are, however, limited in application, and to increase the magnitude of the voltages which may be measured, it is necessary to use a voltage dividing resistor chain (see page 153). To measure voltages less than say 0.1 V, the resolution may be improved by causing the current in the divider circuit to be decreased by a factor of 10 by adding a resistance in series with the divider and simultaneously shunting it with another resistor, as in figure 3.3 where, with the plug switch in position A and the current flowing through the divider (of total resistance R) is $E/R = I$ and when the plug is in position B, the current in the divider becomes

$$i = \frac{I \cdot \dfrac{R}{9}}{R + \dfrac{R}{9}} = \frac{I}{10} \qquad (3.1)$$

The current I through the standardisation resistance Rs remaining constant. This type is known as a double range potentiometer.

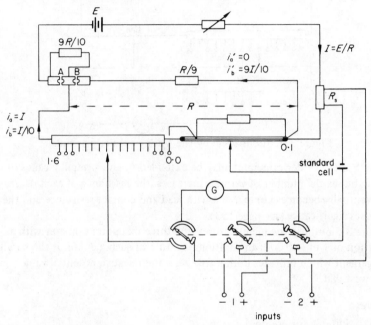

Figure 3.3. Two range potentiometer, with switched standardising and input facility

Two other modifications are illustrated in figure 3.3.

(a)   a fixed resistor R_s having a value adjusted so that with the correct value of current in the divider circuit, the voltage across R_s is equal to the standard cell voltage, and

(b)   the small extension of the slide wire so that zero voltage on it does not coincide with the end stop.

The resolution and hence order of accuracy obtainable for a potentiometer will depend on the degree of subdivision of the resistor units. In the potentiometers so far described the degree of subdivision is limited by the number of steps that it is practicable to construct. To obtain potentiometers with a higher degree of precision, it is necessary to modify further the basic circuit. One of the best methods of obtaining further subdivision of the divider is by means of the Varley vernier principle (originally devised for fault location in submarine cables), which is outlined in figure 3.4 and described in detail in chapter 5. Where (for example) the coarse adjustment would be calibrated in steps of 0.1 V, the vernier in steps of 0.01 V and the slide wire with a total drop of 0.01 V (10 mV) readable to 0.0001 V, (100 μV).

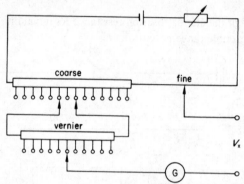

Figure 3.4. Subdivision by the Varley vernier principle

The Varley divider principle[2] may be extended to give several decades of reading but as the number of verniers increases, the resistance of each decreases and eventually becomes comparable with lead and contact resistance and therefore loses significance (see page 153).

Figure 3.5 outlines the main features of a three dial potentiometer with a resolution of 1 in 180 000; an additional feature of such a circuit is the parallel arrangement which provides both a true zero and negative potential values on the lowest dial.

Features

    (a)   Suitable for use on d.c. only.

    (b)   Extremely high input impedance at balance, that is since there is theoretically no current flow between the potentiometer and the source of voltage being measured when they are equal in magnitude the potentiometer presents an infinite input impedance. This does not apply when a voltage divider is being used.

    (c)   Good accuracy is possible, the cheapest potentiometers usually having errors of the order of 0.1 per cent of reading ± 1 unit of the least significant figure, while a good quality instrument has an error of ± 0.001 per cent of reading ± 1 unit of the least significant figure, thus having, in general, errors that are less than those obtainable with the best of the analog pointer instruments, and better than many digital voltmeters (see also page 306).

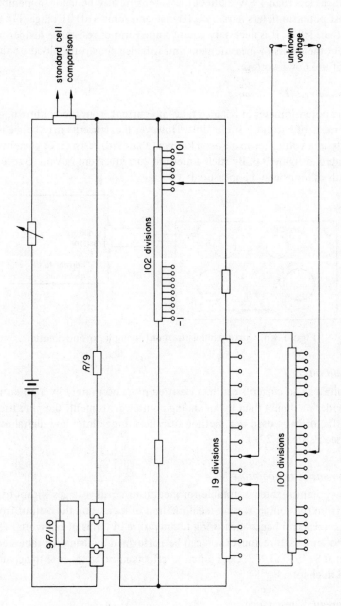

Figure 3.5. Simplified diagram of a three dial potentiometer

standard cell
comparison

unknown
voltage

101

102 divisions

1

R/9

9R/10

19 divisions

100 divisions

Applications

Measurement of d.c. voltages

For voltages less than 1.8 V a direct measurement may be made, remembering that most potentiometers have a ×0.1 range and some a ×0.01 range. For voltages greater than 1.8 V, it is necessary to use some form of resistance divider which will affect the errors of measurement and will also constitute a load on the source of unknown voltage.

Measurement of d.c. currents.

Since the potentiometer at balance takes no current from the unknown, direct measurement of current is impossible. However d.c. currents may be measured indirectly as a voltage drop across a known (standard) resistor, or current shunt as indicated in figure 3.6. By such a method currents from mA up to several thousands of amps may be measured.

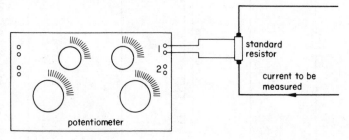

Figure 3.6. Current measurement using a potentiometer

Meter calibration

Since voltages and currents can be measured more accurately by a potentiometer (plus divider or shunt) than by an analog pointer instrument, the potentiometer can satisfactorily be used as a method of calibrating pointer and digital instruments (see also page 221).

Measurement of temperature

Since the potentiometer is suitable for measuring small voltages without taking current from the source, it is an ideal method of measuring the output from thermocouples and hence measuring temperature in terms of millivolts. The voltage to temperature conversion can be performed by using the tables contained in B.S. 4937:1973. Cold junction effects must be taken into account, as outlined in chapter 8.

Measurement of resistance

Low and medium resistances may be measured using a potentiometer provided a standard resistance of a magnitude similar to the unknown and a stable current supply are available. This measurement is performed by measuring first the voltage across the unknown (V_x) and then across the standard (V_s) thus:

$$V_x = R_x I \quad \text{and} \quad V_s = R_s I$$

$$\text{or} \quad R_x = \frac{V_x R_s}{V_s} \tag{3.2}$$

providing the current magnitude is identical for the measurement of both voltage drops. Caution must be exercised in this respect since during the time interval between measuring V_s and V_x, the current may have changed (due to heating effects or supply variations) by a small but sufficiently large amount to affect the errors in measurement. Another precaution which must be taken is to reduce the effect of small thermoelectric e.m.fs occurring at contacts between dissimilar metals. This can be achieved by averaging results for both directions of current in through the resistors.

3.2 A.C. POTENTIOMETERS

Since the potentiometer is such a powerful instrument for the precision measurement of direct voltages, its application to the measurement of alternating voltages is an obvious extension of its utility[2]. There are, however, two difficulties to be overcome; first the measurement of phase, and secondly the standardisation of the potentiometer since a chemical alternating standard cell cannot be made.

The measurement of phase may be performed in one of two ways; either by rotating the measuring voltage until its phase is the same as the unknown and then adjusting the magnitude of the voltage — this type is known as a polar potentiometer — or by using two components of voltage at right angles, their magnitudes being adjusted until the sum of the two components balances with the potential to be measured, this latter form being termed a coordinate potentiometer.

The difficulty of standardising the potentiometer is overcome by using either a current measuring instrument in the potentiometer instead of a standard cell, or by using a transfer instrument, that is a device which measures the heating effect of a d.c. current so that this may be compared with the heating effect of an a.c. current. The value of the d.c. current is standardised by comparison with the voltage from a standard cell (see page 78).

Polar potentiometer

The Drysdale polar a.c. potentiometer (see figure 3.7) was the first complete instrument having a performance comparable with that of a d.c. potentiometer. An essential part of the polar potentiometer is the phase shifting transformer. It consists of a stator wound with two similar windings displaced $90°$ in space with respect to each other. By adjustment of the R and C in figure 3.7 the current in the two windings can be made equal and orthogonal. The phase of the voltage induced in the rotor winding will vary with respect to the supply in accordance with the angular position of the rotor. Thus the phase shifting transformer allows the phase of the current in the potentiometer to be rotated through any angle, while the potentiometer provides variation in magnitude. In this way an unknown voltage can be balanced both for phase and magnitude and the result given as a polar vector.

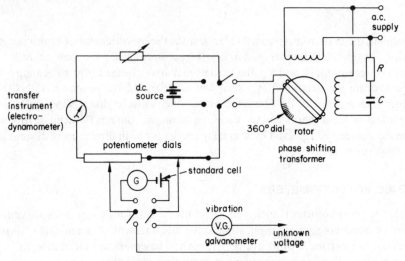

Figure 3.7. Polar a.c. potentiometer

The potentiometer is first standardised on d.c. by energising it with d.c. and balancing against a standard cell. The transfer instrument (electrodynamic milliammeter) will have a full scale reading of about 55 mA and at standardisation the current through this will be 50 mA. After the potentiometer has been standardised on d.c. it is usual to adjust the zero of the milliammeter so that the pointer coincides with a fine standardisation line on the instrument scale.

Since the electrodynamic instrument operates equally well on both d.c. and a.c. (see page 15), if the current flowing when a.c. is applied is adjusted to reproduce exactly the reading obtained during d.c. standardisation, the same effective or r.m.s. value of current must be flowing. If the impedance of the potentiometer coils is the same for a.c. and d.c., then the effective or r.m.s. value of voltage drop along the potentiometer resistance chain will be the same upon a.c. as upon d.c. when standardised in this way.

The current through the electrodynamic milliammeter must be constant at 50 mA for all positions of the phase shifting transformers rotor. If this requirement is satisfied then the potentiometer is calibrated and ready for use.

Coordinate potentiometer

The coordinate a.c. potentiometer[1] consists of two exactly similar potentiometers, one being supplied with current in phase with the supply and the other with current in quadrature with the supply, figure 3.8.

The first stage of the standardising procedure for an a.c. coordinate potentiometer is to standardise the in phase potentiometer on d.c. by comparing the
voltage of a standard cell with the appropriate value set on the potentiometer
dials. The resistor R is adjusted until the galvanometer G indicates a null. On
completion of the d.c. standardisation, the reading of the transfer instrument

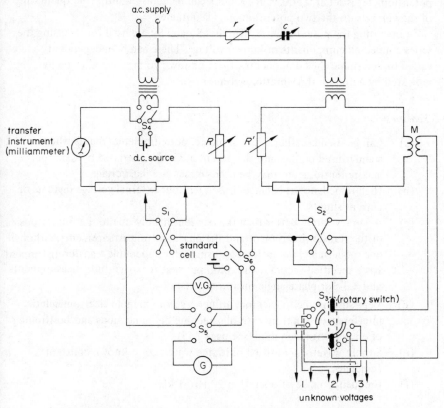

Figure 3.8. Basic circuit of an a.c. coordinate potentiometer

(which may be a torsion head dynamometer milliammeter) is observed. To
facilitate the a.c. standardisation of the in phase potentiometer, it is necessary
to replace the d.c. supply by an a.c. one (S.4), change the light spot galvanometer
for a vibration galvanometer (S.5), and disconnect the standard cell (S.6).
Then by restoring the deflection on the transfer instrument to that recorded
during the d.c. standardisation, the in phase potentiometer is standardised on
a.c.

Now the magnitude of the current in the quadrature potentiometer must be
the same as that in the in phase (normally 50 mA); these two currents must
also be exactly in quadrature. To perform the final stage of the standardisation
procedure, the switch S_3 is put to the position in the diagram. Then since the

e.m.f. induced in the secondary of M will lag $90°$ in phase behind the current in the primary winding (that is in the quadrature potentiometer slide wire) and have a magnitude of 0.5 V (since $V = 2\pi f M i$, and if $M = 0.0318$ H, $f = 50$ Hz and $i = 0.05$ A the induced voltage = 0.5 V). This induced voltage will be in phase with the in phase potentiometer (if correctly in quadrature). So if the in phase potentiometer is set at 0.500 V, R', r and c can be adjusted until true quadrature of the voltages on the two potentiometers is obtained.

In operating the potentiometer, switches S_1 and S_2 are used for reversing the series connected components of known voltage. These can be independently varied to determine the components of an unknown voltage, equality being indicated by a null on the vibration galvanometer.

Features

(a)   Can be used as either a d.c. or an a.c. potentiometer (that is when standardised on d.c. and the potential divider energised by d.c., the 'a.c. potentiometer' may be used to measure d.c. voltages).

(b)   High impedance at balance if used without external voltage dividers or current shunts.

(c)   Accuracy rarely better than 0.2 per cent mainly due to the limitations of the transfer instrument and the phase shifting arrangements, although one modern a.c. potentiometer using a thermocouple transfer instrument has a specified accuracy of ± 0.05 per cent for magnitude measurements and $\pm 3'$ for phase angle measurements.

(d)   Very susceptible to external influences, for example electromagnetic interference, therefore careful screening of connections and positioning of equipment is required (see page 145).

(e)   Supply waveform must be pure sine wave and of known constant frequency.

(f)   Frequency range of operation 20 Hz–1 kHz.

Applications

Measurement of voltage

A.C. voltages up to 1.5 V can be measured directly providing they are of the same frequency as that energising the potentiometer. For voltages greater than this an external voltage divider, suitable for a.c., must be used. The measurement will be in the $V\underline{/\theta}$ form from a polar potentiometer and in the $(a + jb)$ form from a coordinate potentiometer.

Measurement of current

As with the d.c. potentiometer, direct measurement of current is not practicable, but currents may be measured as voltage drops across standard resistors or shunts, the additional requirement for a.c. use being that the resistor has no reactance.

Meter calibration

A use of the above applications is in the calibration of voltmeters and ammeters and this may be extended to the calibration of wattmeters at a variety of power factors. A possible circuit for this application is given in figure 3.9, where the ammeter and voltmeter are included only to maintain a check on the magnitudes

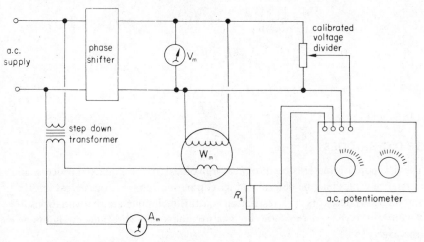

Figure 3.9. Calibration of a wattmeter

of voltage and current supplied to the wattmeter. If the voltage applied to the wattmeter voltage coil is $k(a + jb)$, (k = divider constant) and the current in the wattmeter current coil is $(c + jd)/R_s$, then the effective power applied to the wattmeter is $k/R_s (ac + bd)$ W, and is a satisfactory (if somewhat lengthy) method of calibrating a wattmeter at a low power factor.

Power measurements

From the above, another suitable use of the a.c. potentiometer is the measurement of power and it is particularly suitable for measurements where the total power to be measured is small, for example iron losses of small samples. As in the calibration of a wattmeter, the current is measured as a volt drop across a resistance and the voltage determined either directly or with the aid of a voltage divider. If the voltage can be measured directly, it is advantageous, as under this condition the voltage is measured by an instrument with an infinite input impedance.

Measurement of instrument transformer errors

Instrument transformers (current and voltage scaling devices) have small errors of ratio and introduce a small phase shift between the primary and secondary circuits (see pages 159 and 164). It is necessary that these errors shall be known, and the a.c. potentiometer with its facility to measure magnitude and phase is a possible method of performing these measurements.

Impedance measurements

A determination of impedance requires the measurement of voltage and current magnitudes together with the phase angle between them. If a coordinate potentiometer is used, giving a voltage across the impedance of $V_i + jV_q$ and a current through it of $I_i + jI_q$, then

$$
\begin{aligned}
Z &= \frac{V_i + jV_q}{I_i + jI_q} \\
&= \frac{(V_i + jV_q)(I_i - jI_q)}{I_i^2 + I_q^2} \\
&= \frac{(V_i I_i + V_q I_q) + j(I_i V_q - I_q V_i)}{I_i^2 + I_q^2} \ \Omega
\end{aligned}
\tag{3.3}
$$

3.3 D.C. BRIDGES[3]

The d.c. potentiometer is used to measure voltage by a comparison process, and resistance values may be determined by comparing voltage drops across known and unknown resistors. The restrictions on this technique are that variations in the current in the resistors between measurements can cause a large random error (see page 219).

Wheatstone bridge

If the standard resistance is a known variable, for example a decade resistance box and the potentiometer replaced by a resistance ratio, an arrangement is obtained in which variations in the applied voltage V will theoretically have no

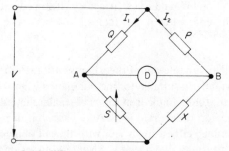

Figure 3.10. Wheatstone bridge (equivalent circuit)

effect on the value found for the unknown. Rearranging such a circuit to that shown in figure 3.10, the equations at balance (zero current through the detector D) are

$$I_1 Q = I_2 P \quad \text{and} \quad I_1 S = I_2 X$$

from which $\quad \dfrac{I_2 X}{I_2 P} = \dfrac{I_1 S}{I_1 Q}$

since X (the unknown) $= \dfrac{SP}{Q}$ Ω (3.4)

and the balance is independent of I_1, I_2 and therefore V, becoming purely a ratio of resistances.

This circuit or bridge arrangement is termed the Wheatstone bridge circuit and is the basis of a considerable amount of instrumentation. In its commercially available form, it is used for the measurement of resistance values in the range 1 Ω to about 11M Ω. Figure 3.11 shows one such circuit. The wise range of the bridge is obtained by having one arm (S) variable from 1 Ω to 10 kΩ and the ratio P/Q adjustable in decade multiples from 0.001 to 1000.

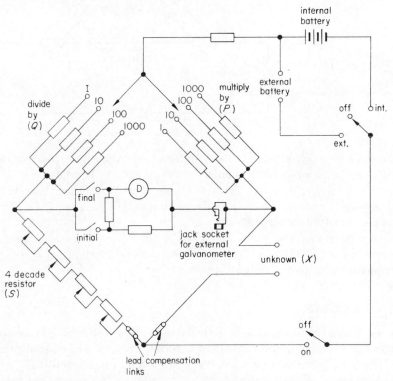

Figure 3.11. Wheatstone bridge, commercial arrangement (Pye–Unicam)

Kelvin double bridge

To measure resistance values less than 1 Ω, a modification of the Wheatstone by Kelvin is very satisfactory. The balance equation for the Kelvin double bridge (figure 3.12) is

$$X = \frac{QS}{M} + \frac{mr}{r + q + m} \left(\frac{Q}{M} - \frac{q}{m} \right) \ \Omega \tag{3.5}$$

In practice $m = M$; $q = Q$ and r (the resistance of the connection between X and S) is made very small and the expression for the value of the unknown becomes:

$$X = \frac{QS}{M} \, \Omega \qquad\qquad (3.6)$$

The advantages of the Kelvin double bridge are that the effects of contact and lead resistances are eliminated and that variations in the current through the

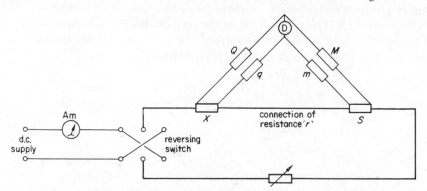

Figure 3.12. Kelvin double bridge

unknown have no effect on the balance of the bridge. Also the resistance of the unknown may be measured at its rated or working current. To reduce the effect of thermoelectric e.m.fs which will occur at junctions of dissimilar metals, it is necessary to reverse the direction of the current through X and S and average the values obtained for the unknown.

The range of resistance values which may be determined with this bridge is normally from 1.000 Ω to 0.100m Ω with an error of 0.1 to 0.01 per cent (depending on the quality of the instrument).

3.4 A.C. BRIDGES

The a.c. bridge circuit, figure 3.13, is similar in principle to the d.c. Wheatstone bridge, but is used to measure capacitance and inductance as well as resistance.

The general principle is the same, that is to obtain a balance so that the detector D gives a null reading when $V_{AE} = V_{AB}$, but with the a.c. bridge these voltages must be equal in magnitude and phase.

Detectors

The source of a.c. is usually an oscillator, the detector being one of the following types:

(a)   *Vibration galvanometer.* A light spot galvanometer which may be tuned to mechanically resonate at the supply frequency — commonly 50 Hz — but may be designed for use up to 1 kHz (see page 12).

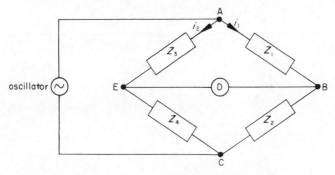

Figure 3.13. The basic a.c. bridge circuit

(b) *Earphones.* Used from about 250 Hz up to 3 or 4 kHz, useful for variable frequency bridges.

(c) *Tuneable amplifier detector*[8]. These are the most versatile of the detectors and consist of a transistor amplifier which may be tuned (electrically) to respond to a narrow bandwidth at the bridge frequency, the amplifier output driving a pointer instrument. This detector being usable over a frequency range of 10 Hz to 100 kHz (see page 32).

The variations of the a.c. bridge circuit are numerous and the details of each are published in many textbooks[1,4,13]. Common examples, figures 3.14 and 3.15, are attributed to Maxwell, Heaviside, Owen, DeSauty, Wien and Schering. The first three are suitable for measuring inductance, the last three for capacitance. Many commercial variations exist on these bridge circuits, but the advent of the transformer bridge (see next section) has tended to oust them. One bridge that is still popular — due to its use in high voltage work — is the Schering bridge[5], and to demonstrate the principles of determining the balance conditions for a.c. bridges, this will be considered in detail.

The balance conditions derived assume a series circuit for the unknown C_1, R_1.

At balance:
$$\left.\begin{array}{c} V_{AB} = V_{AE} \\ \text{and} \quad V_{BC} = V_{EC} \end{array}\right\} \text{ in magnitude and phase}$$

$$\therefore \quad i_1 \left\{ R_1 + \frac{1}{j\omega C_1} \right\} = i_2 \frac{1}{j\omega C_2} \tag{3.7}$$

and
$$i_1 R_3 = i_2 \left\{ \frac{R_4 \cdot \dfrac{1}{j\omega C_4}}{R_4 + \dfrac{1}{j\omega C_4}} \right\}$$

giving
$$i_1 R_3 = i_2 \left\{ \frac{R_4}{1 + j\omega C_4 R_4} \right\} \tag{3.8}$$

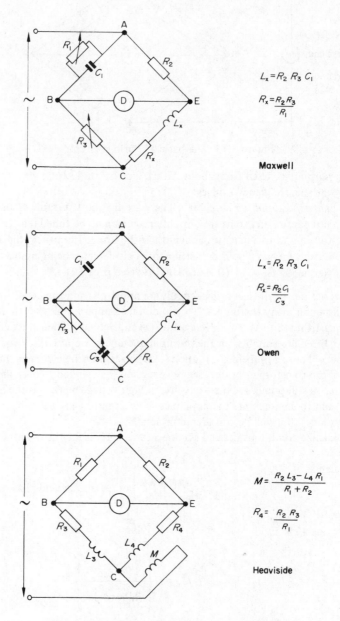

$$L_x = R_2 R_3 C_1$$

$$R_x = \frac{R_2 R_3}{R_1}$$

Maxwell

$$L_x = R_2 R_3 C_1$$

$$R_x = \frac{R_2 C_1}{C_3}$$

Owen

$$M = \frac{R_2 L_3 - L_4 R_1}{R_1 + R_2}$$

$$R_4 = \frac{R_2 R_3}{R_1}$$

Heaviside

Figure 3.14. Some classic inductance bridges

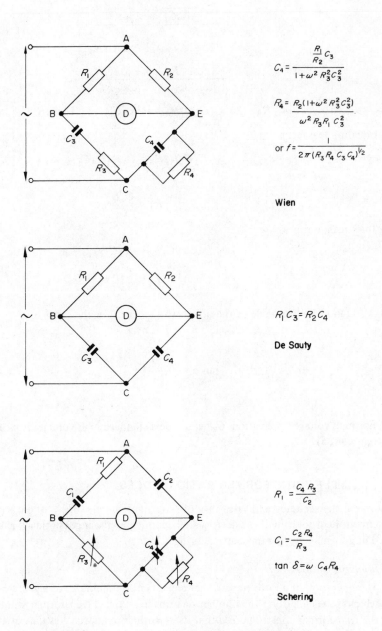

$$C_4 = \frac{\dfrac{R_1}{R_2} C_3}{1 + \omega^2 R_3^2 C_3^2}$$

$$R_4 = \frac{R_2(1 + \omega^2 R_3^2 C_3^2)}{\omega^2 R_3 R_1 C_3^2}$$

or $f = \dfrac{1}{2\pi (R_3 R_4 C_3 C_4)^{1/2}}$

Wien

$$R_1 C_3 = R_2 C_4$$

De Sauty

$$R_1 = \frac{C_4 R_3}{C_2}$$

$$C_1 = \frac{C_2 R_4}{R_3}$$

$$\tan \delta = \omega\, C_4 R_4$$

Schering

Figure 3.15. Some classic capacitance bridges

From (3.7) and (3.8)

$$\frac{R_1 + \dfrac{1}{j\omega C_1}}{R_3} = \frac{\dfrac{1}{j\omega C_2}}{\dfrac{R_4}{1 + j\omega C_4 R_4}}$$

$$\therefore \quad \frac{1 + j\omega C_1 R_1}{j\omega R_3 C_1} = \frac{1 + j\omega C_4 R_4}{j\omega C_2 R_4}$$

$$\therefore \quad j\omega C_2 R_4 - \omega^2 C_2 C_1 R_1 R_4 = j\omega R_3 C_1 - \omega^2 C_4 R_4 R_3 C_1$$

Equating real terms

$$C_2 C_1 R_1 R_4 = C_4 R_4 R_3 C_1$$

$\therefore$

$$R_1 = \frac{C_4 R_3}{C_2} \tag{3.9}$$

Imaginary terms

$$C_2 R_4 = R_3 C_1$$

$$\therefore \quad C_1 = \frac{C_2 R_4}{R_3} \tag{3.10}$$

Also tan δ the loss angle of the unknown capacitor for the series case

$$= \omega C_1 R_1$$

$$\therefore \quad \tan \delta = \frac{\omega C_2 R_4}{R_3} \times \frac{C_4 R_3}{C_2}$$

$$= \omega C_4 R_4 \tag{3.11}$$

For high voltage application, C_2 would be a standard high voltage capacitor (see figure 4.25).

3.5 SIMPLE TRANSFORMER RATIO BRIDGES

Due to the accuracy and versatility of ratio transformers, forms of a.c. bridge circuit incorporating these devices are becoming increasingly popular, and in doing so are eclipsing the classical bridge circuits.

Principle of operation

The operation of a ratio transformer consists of obtaining a voltage division which is dependent on the number of turns in a tapped transformer winding, the transformer constructed so that its performance approaches that of an 'ideal' transformer, (that is no core loss, windings with perfect coupling and zero resistance). Then the voltage E appearing at the terminals of a transformer winding of N turns is:

$$E = 4KN\Phi_m f$$

where K is a constant depending on the waveform of the flux, which has a maximum value of Φ_m and f is the frequency of flux variations. Thus for given values of Φ_m, f and K,

$$E = K'N \qquad (3.12)$$

Figure 3.16. Tapped autotransformer

If the transformer in question is of the autotransformer type (figure 3.16), the division of the applied voltage E into E_1 and E_2 will be:

$$E_1 = \frac{E \cdot N_1}{N} \quad \text{and} \quad E_2 = \frac{E \cdot N_2}{N}$$

assuming the transformer is ideal.

In practice the 'ideal' transformer cannot be constructed, but the ideals of perfect coupling, zero winding resistance and zero core loss can be approached. Consider the core loss; this may be reduced to negligible proportions by, first, selecting a core material which has the smallest magnetic hysteresis and eddy current losses possible at the intended frequency of operation, and second, constructing the core to have a minimum reluctance (that is absence of air gaps and use of the core material in its preferred grain direction, resulting in a toroidal

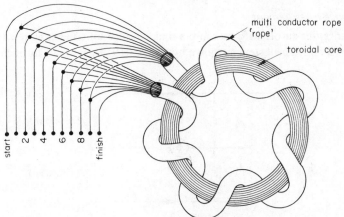

Figure 3.17. Tapped toroidal transformer winding

clock spring core). To approach the ideal of perfect coupling requires that the magnetic flux produced by one turn should link all the others, that is a transformer with zero leakage flux. Fortunately a winding on toroidal core has a very small leakage reactance, and this can be further reduced by using a high permeability core material and windings which take the form of a multiconductor rope (figure 3.17). If the multiconductor rope has ten wires, connecting successive sets of turns in series and taking a tapping from each joint to a terminal, could provide a decade of voltage division.

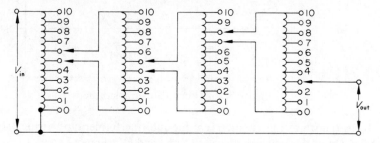

Figure 3.18. Four decade ratio transformer, set to 0.5473

There remains only the problem of reducing the series resistance of the winding, the possibilities here are to use as large a copper cross-section as possible, and to use a core material such that the transformer has as high a Q as possible. Using the techniques outlined above, an inexpensive 4 decade ratio transformer with a ratio error less than 1 part in 10^4 can be constructed. The successive decades being derived from an arrangement similar to a Kelvin Varley divider (figure 3.18 – see also page 153).

Applications

Resistance measurement

To arrange this autotransformer into a simple bridge for comparing resistance values is a fairly small step, as shown by figure 3.19. The current through R_x is

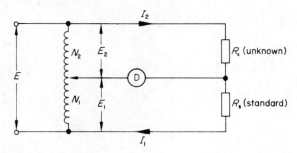

Figure 3.19. Resistance measurement using a ratio transformer

$$I_2 = \frac{E_2}{R_x} = \frac{K' N_2}{R_x}$$

and the current through R_s is

$$I_1 = \frac{E_1}{R_s} = \frac{K' N_1}{R_s}$$

At balance the difference between these currents will be zero, that is $I_1 = I_2$ and

$$\frac{K' N_2}{R_x} = \frac{N_1 K'}{R_s}$$

or

$$R_x = R_s \cdot \frac{N_2}{N_1} \tag{3.13}$$

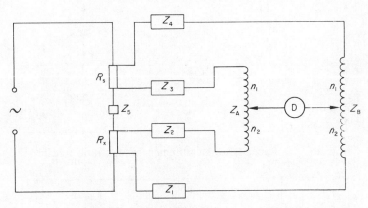

Figure 3.20. Measurement of a low resistance

By using two ratio transformers a form of Kelvin double bridge for measurements of low resistances may be devised (figure 3.20), where

$$R_x = R_s \left[\frac{n_2 Z_B + Z_1}{(1 - n_2) Z_B + Z_4} \right] + Z_5 \left[\frac{(1 - n_1) Z_A + Z_3}{Z_1 + Z_2 + Z_3 + Z_5} \right] \times$$
$$\left[\frac{n_2 Z_B + Z_1}{(1 - n_2) Z_B + Z_4} - \frac{n_1 Z_A + Z_2}{(1 - n_1) Z_A + Z_3} \right] \tag{3.14}$$

and this expression[10] may be simplified to

$$R_x = \frac{R_s \cdot n_2}{n_1} \tag{3.15}$$

(providing lead impedances are small and R_s and R_x are of the same order of magnitude).

This form of Kelvin double bridge can, of course, only be used on a.c. but by plotting R_x against frequency, an extrapolation for the d.c. value of R_x may be performed to determine R_x within a few parts per million.

Capacitance measurement

For the measurement of capacitance a circuit of the form in figure 3.21 can be used, when

$$C_x = \frac{C_s N_1}{N_2} \tag{3.16}$$

and the parallel leakage of the resistance of the capacitor

$$R_x = \frac{R_s N_2}{N_1}$$

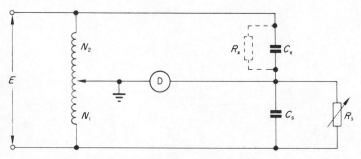

Figure 3.21. Capacitance measurement using a ratio transformer

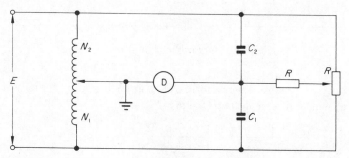

Figure 3.22. A circuit for the measurement of a 'good' capacitor

An alternative arrangement of this circuit, sometimes used in connection with capacitance transducers (see page 254), is shown in figure 3.22 where providing the parallel leakage resistance of both capacitors is large, the resistance effects can be neglected and

$$C_2 = \frac{C_1 N_1}{N_2}$$

Phase angle measurement

For measurements of small phase angles[11,12] the circuits in figures 3.23a and b may be used. The resistance R must be large to minimise loading on the ratio transformer and a tuned detector used so that only unbalance in the fundamental frequency is indicated.

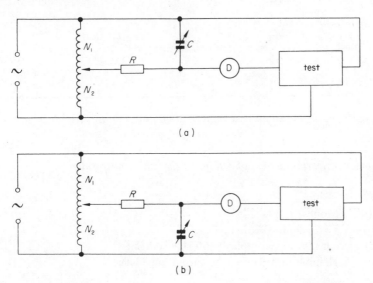

(a)

(b)

Figure 3.23. Phase angle measurement circuits (a) for negative (b) for positive phase angles

The ratio transformer and the capacitor being alternately adjusted until balance is obtained.

It may then be shown that the phase angle $\phi = \tan^{-1} (-\omega RC)$ and the magnitude of the in phase component is

$$\frac{N_2}{N_1 + N_2} \times \cos^2 \phi \qquad (3.17)$$

Conversion of fixed standard into a variable one

Another application of the ratio transformer[11,12] is as a means of converting a fixed standard into a variable one, and this has applications in checking certain bridge circuits.

Other Uses

Other application examples of the auto or ratio transformer are in calibration of meters and in the determination of amplifier gain and transformer ratios.

In practice the ratio transformer is not ideal and in all its uses which involve drawing a load current through part of the winding, allowance should be made for the resistance and inductance of the windings. The equivalent circuit which represents the ratio transformer is shown in figure 3.24. The values of R_s and L_s will vary with ratio setting and the error they introduce must be added to the no-load error which is the value normally quoted.

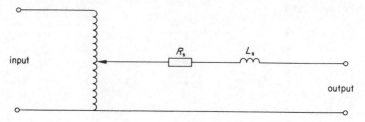

Figure 3.24. Simplified equivalent circuit of a ratio transformer

Summary of features and applications of the ratio transformer

Features

(a)   a.c. only;
(b)   very small ratio errors;
(c)   frequency range, typically 50 Hz—50 kHz;
(d)   high input impedance, low output impedance.

Applications

(a)   Measurement of resistance by comparison with a standard resistance;
(b)   measurement of capacitance by comparison with a standard capacitance;
(c)   measurement of inductance by comparison with a standard inductance;
(d)   measurement of amplifier gain and phase shift;
(e)   may be used as a voltage divider.

3.6 TRANSFORMER DOUBLE RATIO BRIDGES

The transformer bridges described so far have had a similarity with the classical bridges in that at balance, a voltage equality exists and there is no resultant current flow through the detector. An alternative to this would be to arrange that, at balance, the current flowing through the unknown was equal and opposite to that through the standard or known impedance, the detector indicating this equality.

Considering the circuit in figure 3.25 where the 'ideal' voltage transformer has a secondary winding of N_1 turns with a tapping at N_2 turns, so that voltages E_1 and E_2 are applied to impedances Z_x and Z_s respectively, resulting in currents I_1 and I_2 flowing in the windings n_1 and n_2 of the current comparator.

The ampere turns $n_1 I_1$ and $n_2 I_2$ will produce fluxes in the core of the comparator that will oppose each other and a balance condition will be attained when the resultant flux is zero, this being indicated by the voltage in the winding connected to the detector also being zero.

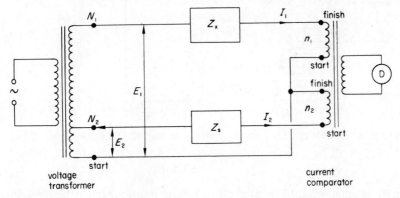

Figure 3.25. Basic circuit of a double ratio transformer bridge

Thus at balance $I_1 n_1 = I_2 n_2$ (3.18)

Now $I_1 = E_1/Z_x = kN_1/Z_x$; and $I_2 = E_2/Z_s = kN_2/Z_s$

$\therefore \quad \dfrac{N_1 n_1}{Z_x} = \dfrac{N_2 n_2}{Z_s}$ or $Z_x = Z_s \cdot \dfrac{N_1 n_1}{N_2 n_2}$ (3.19)

If Z_x is resistive,

$$R_x = R_s \cdot \frac{N_1 n_1}{N_2 n_2}$$ (3.20)

If Z_x is capacitive,

$$\frac{1}{j\omega C_x} = \frac{1}{j\omega C_s} \cdot \frac{N_1 n_1}{N_2 n_2}$$

or

$$C_x = C_s \cdot \frac{N_2 n_2}{N_1 n_1}$$ (3.21)

The circuit described above requires a standard that is capable of being varied in value, and if a reasonable degree of accuracy is desired it would be an expensive arrangement. However if the voltage transformer is tapped as shown in figure 3.26, C_s could have a fixed value and the $N_2 C_s$ product be the variable standard.

Reconsidering the equation (3.20) it may be made to have a similar form to (3.21) by converting resistances to conductances, that is

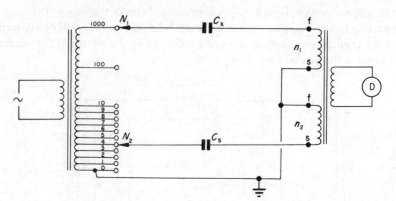

Figure 3.26. Double ratio transformer bridge having a fixed value standard (C_s)

$$\frac{1}{R_x} = G_x = G_s \cdot \frac{N_2 n_2}{N_1 n_1}$$

Then if the unknown is a capacitor with a leakage resistance it may be represented as an admittance $Y_x = G_x + j\omega C_x$ that can be balanced against conductance and capacitance standards G_s and C_s as shown in figure 3.27 for which:

$$G_x + j\omega C_x = \frac{n_2}{N_1 n_1} \left\{ G_{s1}N_{21} + G_{s2}N_{22} + j\omega \left(C_{s1}N'_{21} + C_{s2}N'_{22} \right) \right\} \qquad (3.22)$$

Only two decades of standard have been used for the bridge shown in figure 3.27: more could have been used although this tends to make the switching arrangements

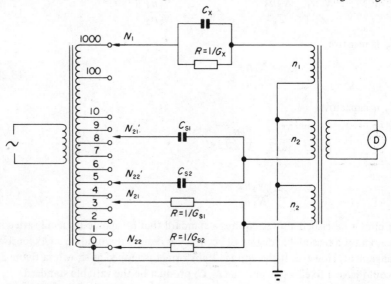

Figure 3.27. Double ratio transformer bridge having fixed value capacitance and conductance standards

complicated. A compromise between a large number of switched standards and expensive variable standards is employed in most commercial bridges, resulting in a circuit of the form shown in figure 3.28 where in addition to the switched standards, a variable capacitance standard (C_{s3}) is permanently connected to the 10-turn tapping of the voltage transformer, while a variable conductance standard

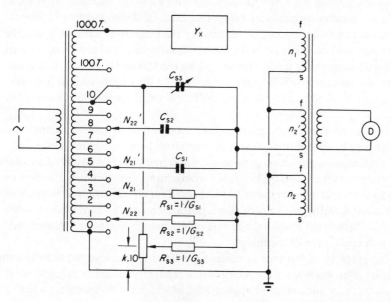

Figure 3.28. Double ratio transformer bridge arranged for capacitance measurement

(G_{s3}), which theoretically may be required to have a value that can be varied from 0 to 10 μS (∞ to 100 kΩ), is in practice likely to be a 100 kΩ resistor connected to a continuously variable resistive divider that is connected across 10 turns of the voltage transformer; the travel of the wiper being calibrated as fractions of the 10 turns.

3.6.1 Capacitance measurement

The circuit in figure 3.28 is a double ratio transformer bridge arranged for capacitance measurement. The unknown is a capacitor having parallel components R_x and C_x, and the equation of balance will be:

$$\frac{1}{R_x} + \frac{1}{} = \frac{1}{N_1 n_1} \left\{ n_2 (N_{21} G_{31} + N_{22} G_{s2} + k10\, G_{s3}) + \right.$$

$$\left. j\omega n_2' (N_{21}' C_{s1} + N_{22}' C_{s2} + 10 C_{s3}) \right\} \quad (3.23)$$

If in the bridge represented by figure 3.28

$$C_{s1} = 100 \text{ nF} \quad G_{s1} = 1 \text{ mS} \quad n_1 = 100 \text{ turns}$$
$$C_{s2} = 10 \text{ nF} \quad G_{s2} = 100 \, \mu\text{S} \quad n_2 = 100 \text{ turns}$$
$$C_{s3} = 630 \text{ pF} \quad G_{s3} = 10 \, \mu\text{S} \quad n_2' = 1000 \text{ turns.}$$

and k (the fraction of the resistive divider to which G_{s3} is connected) is 0.35. Substituting these values in equation 3.23 gives $1/R_x = 3.135 \, \mu\text{S}$ and $C_x = 5.863 \text{ nF}$ thus the unknown is a 5.863 nF capacitor with a leakage resistance of 319 k Ω.

In a commercially produced bridge (for example those marketed by Wayne Kerr) the turns values would be adjusted by the range selection switch, and the display at balance would be in the form of conductance and capacitance values, each with appropriate units. Thus in making measurements on a capacitor, the value of capacitance would be read directly from the display and the leakage resistance determined by computing the reciprocal of the displayed conductance value.

The standards of conductance and capacitance that are used in a transformer bridge may have impurities, that is the capacitors have leakage resistance, and the conductances a reactive component. These impurities are balanced out during the manufacture of a bridge by circuit arrangements that permit the ampere turns due to the impurity component to be opposed for all conditions of bridge operation. A further refinement that is normally incorporated is a zero balance or trim control that enables the effects of lead capacitance to be balanced out for each application of the bridge.

The property of this type of bridge that allows the resistive and reactive components of an unknown to be adjusted independently of each other facilitates its use in a very wide range of applications, of which the following are a few examples.

3.6.2 Measurement of inductance

So that the independent balance of the resistive and reactive ampere turns can be achieved any inductive circuit must be considered as a two terminal network, the components of which are in parallel. The ampere turns due to the resistive component of the unknown may then be opposed by those through the conductance standard, while those resulting from the inductive component of the unknown must be equated with the ampere turns due to the capacitive standard. To obtain a balance of these latter quantities it is necessary to reverse the current direction through the winding connected to the capacitance standard. This condition of bridge operation normally being indicated by the addition of a negative sign to the display of capacitance value (C_m).

Thus $1/R_p + 1/j\omega L_p = G_m - j\omega C_m$ and since it is usually the series components of an inductor that are required, these may be computed by a parallel to series conversion that results in:

$$\text{series resistance } R = \frac{G_m}{G_m^2 + \omega^2 C_m^2} \tag{3.24}$$

and

$$\text{series inductance } L = \frac{C_m}{G_m^2 + \omega^2 C_m^2} \tag{3.25}$$

3.6.3 Measurement of low impedance

The range of a double ratio transformer bridge may be extended[14,15] to measure low value impedances by the use of a pair of known noninductive resistors each

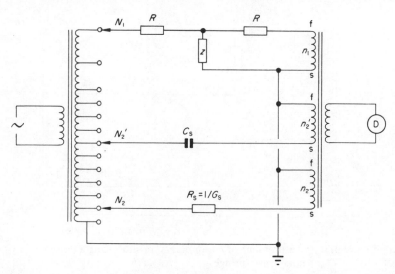

Figure 3.29. Double ratio transformer bridge adapted for low impedance measurement

having a value R and connected with the unknown (z) to form a T network as shown in figure 3.29. Then providing $R \gg z$,

$$z = \frac{R^2}{N_1 n_1} \left\{ N_2 n_2 G_s + j\omega C_s N_2' n_2' \right\}$$

or for a calibrated bridge

$$z = R^2 (G_m + j\omega C_m) \tag{3.26}$$

3.6.4 Measurements of components 'in situ'

One of the greatest advantages of double ratio transformer bridges is the ability to use them for the measurement of components whilst these remain connected in a circuit. To appreciate how this is possible it is necessary to consider the transformers that are used in the construction of the double ratio bridge. If these transformers were 'ideal', that is if they had zero leakage reactance, zero winding resistance, and zero magnetisation loss, the impedances Z_e and Z_i in figure 3.30 would be zero, and the voltage applied to Z_u (the unknown) would be unaffected by current flowing through Z_x. In addition to this all the current flowing through Z_u would flow into the n_1 winding, since $I_y = (Z_i . I_u)/(Z_i + Z_y)$. In practice Z_e and Z_i will have small finite values and a correction must be applied to the values read from the balanced bridge. It may be shown that to a close approximation the true value of the unknown impedance[15]

$$Z_u = \left\{ \frac{1}{G_m + j\omega C_m} \right\} \left\{ 1 - \left[\frac{Z_e}{Z_x} + \frac{Z_i}{Z_y} \right] \right\} \qquad (3.27)$$

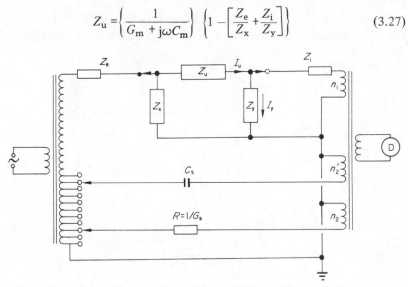

Figure 3.30. Measurement of components 'in situ' using a double ratio transformer bridge

The impedances Z_e and Z_i must be determined (for a particular range of the bridge) by measurements on a separate delta network that contains known values for Z_x and Z_y.

3.6.5 Network characteristics

As the conductance and capacitance standards may be made effectively positive or negative merely by reversing the current direction in part of the current comparator, the double ratio bridge is extremely useful for measuring network characteristics. An example of this is illustrated by figure 3.31 which shows an arrangement that may be used to determine the gain and phase shift of an amplifier that requires a terminating resistor R_t. When the bridge is balanced,

$$\text{gain} = R_t \ [G_m^2 + \omega^2 \ C_m^2]^{\frac{1}{2}}$$
$$\text{and phase shift} = \tan^{-1} \ (\omega C_m)/(G_m).$$

This technique may be used for a number of applications, examples being the determination of (a) the ratio and phase shift of a transformer, and (b) the parameters of a transistor.

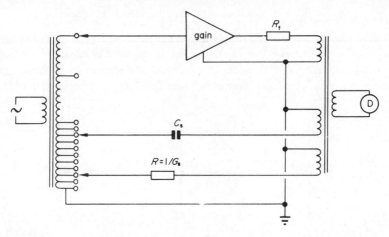

Figure 3.31. Circuit arrangement for the measurement of network characteristics by a double ratio transformer bridge

3.6.6 Comparators

The technique of ampere turn comparison used in the multiratio bridges has been exploited to produce some very high quality 'bridges' known as comparators, the applications of these devices[4-7] being in Standards Laboratories for the determination of current transformer errors, measurement of resistance and d.c. voltage to accuracies of 1 part in 10^6.

Summary of double ratio bridge features

   (a)   a.c. use only;
   (b)   each bridge designed for use over a wide frequency range (see page 115);
   (c)   a 'universal' bridge, that is measurement of R, L, C and network characteristics;
   (d)   can measure components 'in situ' and therefore unaffected by lead capacitance, etc.:
   (e)   good accuracy and resolution obtainable without prohibitive cost;
   (f)   balance is dependent on equality of ampere turns in the current comparator winding.

3.7 TWIN 'T' NETWORKS

These can have zero transmission (with all components finite) and can thus be used in a similar manner to a bridge, that is the balance conditions may be calculated, an impedance to be measured introduced into one of the network arms and the other arms adjusted for balance.

Principle of operation

Consider an a.c. voltage (V_{in}) applied to the input in figure 3.32. Zero output will occur when the currents due to each of the *T*s are equal in magnitude but opposite in phase, that is $I + I_1 = 0$. Under this condition the impedance of the detector does not matter, providing it is not infinite. This fact can be utilised in obtaining

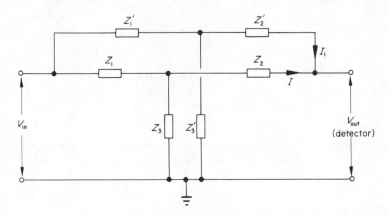

Figure 3.32. Twin 'T' network

a solution for the balance condition, for the network containing $Z_1\ Z_2\ Z_3$ may be considered in isolation, having an input impedance of:

$$Z = Z_1 + \frac{Z_2 \cdot Z_3}{Z_2 + Z_3} \qquad (3.28)$$

and drawing a current from the source of:

$$\frac{V_{in}(Z_2 + Z_3)}{Z_1 Z_2 + Z_2 Z_3 + Z_1 Z_3} \qquad (3.29)$$

and I, the current flowing from the $Z_1\ Z_2\ Z_3$ network to the detector will be:

$$\frac{Z_3}{(Z_2 + Z_3)} \cdot \frac{V_{in}(Z_2 + Z_3)}{(Z_1 Z_2 + Z_2 Z_3 + Z_3 Z_1)}$$

which for balance must be equal and opposite to the current I_1, that is

$$\frac{Z_3 \cdot V_{in}}{Z_1 Z_2 + Z_2 Z_3 + Z_3 Z_1} + \frac{Z_3' \cdot V_{in}}{Z_1' Z_2' + Z_2' Z_3' + Z_3' Z_1'} \qquad (3.30)$$

and since $V_{in} \neq 0$

$$\frac{Z_1' Z_2'}{Z'_3} + Z_2' + Z_1' + \frac{Z_1 Z_2}{Z_3} + Z_2 + Z_1 = 0 \qquad (3.31)$$

and is the condition for balance.

Features

(a)   Both detector and source may have one side earthed, making shielding and earthing easier than in a bridge.

(b)   May be used over the audio to high frequency range as a bridge or rejection filter.

Application

Frequency measurement

For the circuit in figure 3.33a

$$Z_1 = \frac{1}{j\omega C_1} = Z_2 \qquad Z_3 = R_1$$

$$Z_1' = Z_2' = R_2 \qquad Z_3' = \frac{1}{j\omega C_2}$$

$$\therefore \quad \frac{-1}{\omega^2 C_1^2 R_1} + \frac{2}{j\omega C_1} + R_2^2\, j\omega C_2 + 2R_2 = 0$$

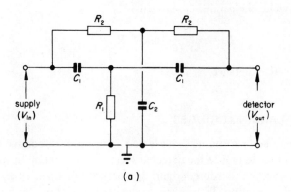

(a)

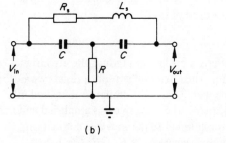

(b)

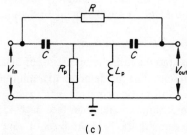

(c)

Figure 3.33. Uses of twin and bridged 'T' networks. (a) Frequency sensitive twin 'T'; (b) measurement of series resistance and inductance

Taking the real part,

$$2R_2 = \frac{1}{\omega^2 C_1^2 R_1}$$

and imaginary part,

$$\frac{2}{\omega C_1} = R_2^2 \, \omega C_2$$

If $R_1 = \dfrac{R_2}{2}$ and $C_1 = \dfrac{C_2}{2}$

both equations give

$$\omega^2 = \frac{1}{C_1^2 R_2^2} \quad \text{or} \quad \omega = \frac{1}{C_1 R_2} \tag{3.32}$$

Bridged T

This is a special case of the twin T network in which $Z_3' = \infty$ and $Z_1 + Z_2 = Z_4$ and the balance equation becomes:

$$\frac{Z_1 \cdot Z_2}{Z_3} + Z_1 + Z_2 + Z_4 = 0 \tag{3.33}$$

figure 3.33b shows the use of this circuit for measuring the series components of an inductance where $L_s = 2/(\omega^2 C)$ and $R_s = 1/(\omega^2 C^2 R)$ and has the disadvantage of the unknown being isolated from earth (shielding problems). The circuit in figure 3.33c yields the parallel components

$$L_p = \frac{1}{2\omega^2 C} \quad \text{and} \quad R_p = \frac{1}{\omega^2 C^2 R} \tag{3.34}$$

for an unknown inductance.

3.8 SELF BALANCING BRIDGES

The principle of comparison methods is that an operator is required to perform manual operations to reduce the detected signal to a minimum. In some instances this can be a lengthy procedure, requiring experience of a particular type of bridge to obtain rapid operation. Thus in circumstances that dictate a bridge measurement as the only satisfactory method and many measurements are to be made, automatic or self balancing bridges have been devised[9].

Principle of operation

As stated above the operation of balancing a bridge is to reduce the error signal presented to the detector to a minimum. Thus in a bridge using a resistance ratio the automatic operation is fairly simple, being basically similar to that used in a potentiometric recorder (see page 45), that is the error signal is amplified and used to energise the field windings of a motor coupled to the movable contact on a resistance divider, for example figure 3.34 where $R_x = R_s \, . \, (R_2/R_1)$.

If R_s has a value of (say) 1000.0 Ω and the movable contact is connected via gearing to a numerical indicator covering a range of 0.0–999.9, the value

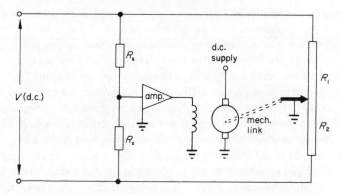

Figure 3.34. Self balancing d.c. Wheatstone bridge

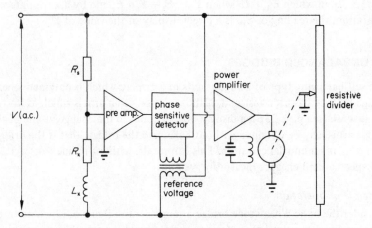

Figure 3.35. Self balanceing a.c. bridge

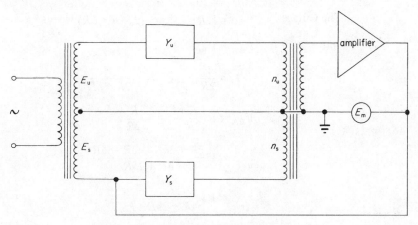

Figure 3.36. Self balancing transformer bridge

of R_x may be automatically displayed using the mechanical position of the contact as a multiplication ratio on the value of R_s.

The above is a simple example in which the energising voltage is d.c. As with a potentiometric pen recorder if the error signals are small the problems of amplifier drift are overcome by using a chopper amplifier of either the electromagnetic type or photoresistive type. If, however, the energising voltage were a.c. and R_x had a reactive component, the bridge would lose sensitivity, but this could be restored by inserting a phase sensitive detector (see page 32) between the preamplifier and power amplifier stages, as in figure 3.35.

Another form of self balancing bridge[9,16] is one in which the error signal supplies an amplifier using feedback to restore balance in a transformer bridge. The ratio of the unknown to the standard being directly proportional to the magnitude of the feedback. A schematic diagram of such a circuit is given in figure 3.36, in which $E_m = E_s$ when $Y_u n_u E_s = Y_s n_s E_s$ and by appropriate calibration, the reading of E_m is a direct display of the value of Y_u.

3.9 UNBALANCED BRIDGES

The self balancing type of bridge results in a display which is obtained when bridge balance has been restored, if the variation of an arm is small, as in strain gauge work (see page 236), a display proportional to the change in one of the bridge arms may be obtained without balancing the bridge, that is the bridge is operating in an unbalanced state. This can result, without undue loss of accuracy, in a speedier and cheaper measuring system.

Principle of operation

Consider the basic Wheatstone bridge arrangement in figure 3.37 where three arms are fixed value resistors of resistance R Ω and the fourth arm variable by a small amount δR.

$$\text{The voltage} \quad V_{AC} = I_1 R; \quad V_{BC} = I_2 (R + \delta R)$$

$$\text{and} \quad V_{BA} = V_{BC} - V_{AC}$$

$$\text{Now} \quad I_1 = \frac{V}{2R} \quad \text{and} \quad I_2 = \frac{V}{2R + \delta R}$$

$$\therefore \quad V_{BA} = V\left\{\frac{R + \delta R}{2R + \delta R} - \frac{R}{2R}\right\}$$

$$= V\frac{2R + 2\delta R - 2R - \delta R}{4R + 2\delta R}$$

$$= V \cdot \frac{\delta R}{4R + 2\delta R} \tag{3.35}$$

This expression indicates that providing δR is small, $V_{AB} \propto \delta R$, and if δR is less than 10 per cent of R, this assumption of proportionality is justifiable, that is $V_{AB} = V(\delta R/4R)$. The current through the detector will depend on its impedance and the total equivalent impedance between the detector and the voltage source, that is the impedance the detector 'sees'. This impedance is known

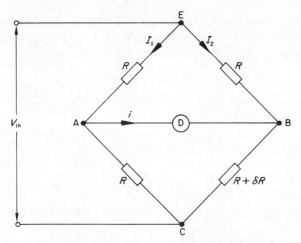

Figure 3.37. Unbalanced Wheatstone bridge

as the Thévenin impedance Z_{th} and is calculated by replacing the source by its internal impedance and computing the impedance of the resultant network, figure 3.38.

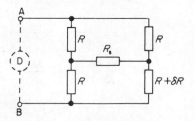

Figure 3.38. Rearrangement of unbalanced Wheatstone bridge (detector open circuit)

In general the impedance of the voltage source used in conjunction with such a circuit will be very small and R_s may be taken as zero. This simplifies the calculation of Z_{th} considerably

$$Z_{th} = \frac{R}{2} + \frac{R(R + \delta R)}{2R + \delta R} \qquad (3.36)$$

and if $\delta R \ll R$

$$Z_{th} \approx \frac{R}{2} + \frac{R}{2} = R$$

and the circuit may be redrawn as figure 3.39, where R_g is the internal resistance of the detector, hence the detector current

$$i_g = \frac{V_{AB}}{Z_{th} + R_g} \qquad (3.37)$$

(V_{AB} being the voltage which would occur at the AB terminals if the detector were open circuit). Such a circuit will produce only small output voltages which can be used to drive a sensitive pointer instrument or pen recorder movement.

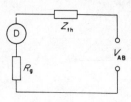

Figure 3.39. Equivalent circuit of figure 3.38

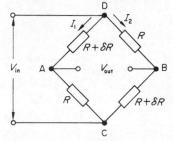

Figure 3.40. Unbalanced bridge with two variable arms

The output of such a system can be increased by using more than one variable arm, for example as shown in figure 3.40, from which

$$V_{AC} = I_1 R; \qquad V_{BC} = I_2 (R + \delta R)$$

$$I_1 = \frac{V}{2R + \delta R}; \qquad I_2 = \frac{V}{2R + \delta R}$$

$$V_{AB} = V \left\{ \frac{R + \delta R}{2R + \delta R} - \frac{R}{2R + \delta R} \right\}$$

$$= V \times \frac{\delta R}{2R + \delta R} \approx V \frac{\delta R}{2R} \qquad (3.38)$$

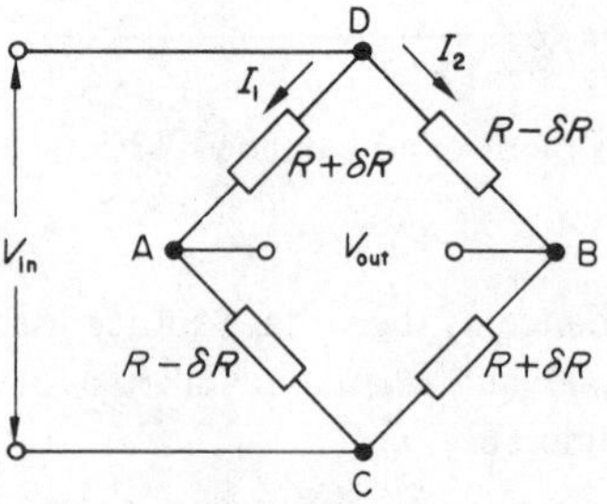

Figure 3.41. Unbalanced bridge circuit with four variable arms

and finally figure 3.41, where the four arms are variable as shown giving

$$V_{AC} = I_1 (R - \delta R)$$

Table 3.1. Comparison table of null (comparison) methods

Type	Errors as percent of reading	Resolution	Range — Resistance	Range — Inductance	Range — Capacitance	d.c.	Frequency (Hz)	Remarks
d.c. potentiometer	0.002 → 0.1	5 p.p.m.	0.01 → 1000Ω			▨ (d.c.)		by comparison with standard
a.c. potentiometer	0.2 → 1	1 part in 10^3	yes	yes	yes (by measuring voltage and current)		▨ at ~100	ditto; voltage levels are criteria for magnitudes of impedance
Wheatstone bridge	0.002 → 1	10 p.p.m.	1Ω → 1MΩ			▨ (d.c.)		may use balanced or unbalanced mode of operation
Kelvin double bridge	0.002 → 1	10μΩ	1mΩ → 1Ω			▨ (d.c.)		unaffected by current variations
Classic a.c. bridges	0.01 → 1	1 part in 10^4	yes	yes	yes		▨ 100 → 10M	see references e.g. Hague
Transformer ratio bridge	0.001 → 0.05 plus error in standard	1 part in 10^7	yes	yes	yes		▨ 10K → 100K; ▨ 100K → 1M	by comparison with standard
Transformer double ratio bridge	0.01 → 0.1 → 1	1 part in 10^5	yes	yes	yes		▨ 100 → 1K; ▨ 100K → 10M; ▨ 10M → 100M; ▨ 100M → 1G	plus 'in situ' measurements
Q meter	2 → 10	1 part in 10^3	impurity component	yes	yes		▨ 10K → 1G	

$$V_{BC} = I_2 (R + \delta R)$$

$$I_1 = \frac{V}{2R} ; \quad I_2 = \frac{V}{2R}$$

$$\therefore \quad V_{AB} = \left\{ \frac{R + \delta R}{2R} - \frac{(R - \delta R)}{2R} \right\} V = V \cdot \frac{\delta R}{R} \tag{3.39}$$

This last circuit is one commonly used in strain gauge transducers (see page 244).

Features

(a)   May be used in either a.c. or d.c. circuits but the operation is simpler in d.c. circuits.
(b)   Only suitable for small variations in component values.
(c)   The output from such a circuit can be directly connected to a suitable recorder.

Applications

(a)   Extensive transducer applications, for example measurement of strain, force, pressure and displacement.
(b)   Measurement of temperature by variations in resistance.
(c)   Any application where it is desirable to convert a variation of resistance into a variation of voltage for recording purposes.

3.10 COMPARISONS

Table 3.1 presents a comparison table for the devices described in this chapter. M. B. Stout[13] summarises the details of the classical a.c. bridges.

REFERENCES

1   E. W. Golding and F. C. Widdis. *Electrical Measurements and Measuring Instruments*. Pitman, London (1963)
2   D. C. Gall. *Direct and Alternating Current Potentiometer Measurements*. Chapman and Hall, London (1938)
3   D. S. Luppold. *Precision D.C. Measurements and Standards*. Addison Wesley, Reading, Mass (1969)
4   B. Hauge and T. R. Foord. *A.C. Bridge Methods* (6th edn.). Pitman, London (1971)
5   N. L. Kusters and O. Peterson. 'A Transformer Ratio Arm Bridge for High Voltage Capacitance Measurements'. *I.E.E.E. Trans. on Communication and Electronics.* 82, pt 1, 606–611 (1963)

6   M. P. MacMartin and N. L. Kusters. 'A Direct Current Comparitor Ratio Bridge for 4-terminal Resistance Measurements'. *I.E.E.E. Trans. of Instrumentation and Measurement.* **IM–15** (1966)

7   J. Sutcliffe. 'Current Comparitor Bridges and Potentiometers'. *Electronic Equipment News* (May 1970)

8   S. D. Prensky. *Electronic Instrumentation.* Prentice-Hall, Englewood Cliffs, N.J. (1963)

9   R. Calvert and J. Mildwater. 'Self-balancing Transformer Ratio Arm Bridges'. *Electronic Engineering.* **35**, No. 430 (Dec. 1963)

10   J. J. Hill and A. P. Miller. 'An a.c. Double Bridge with Inductively Coupled Ratio Arms for Precision Platinum Resistance Thermometry'. *Proc. I.E.E.* **110**, No. 2 (1963)

11   A. H. Silcocks. 'Revolution in Calibration'. *Cambridge Technical Review* (May 1968)

12   A. H. Silcocks. 'Measurement Techniques'. *Kent Technical Review* (March 1971)

13   M. B. Stout. *Basic Electrical Measurements* (2nd edn.). Prentice-Hall, Englewood Cliffs, N.J. (1960)

14   K. Fletcher. 'Use of Transformer Ratio Arm Bridges in Component Measurements'. *Radio and Electronic Components.* 117–123 (Feb. 1962)

15   R. Calvert. 'The Transformer Ratio-Arm Bridge'. *Wayne Kerr Monograph No. 1*

16   The General Radio Company. *The Experimenter.* **39**, No. 4 and **40**, nos 11, 12.

4

Interference and Screening

In practical engineering the conduction of an electrical signal from the measurand to a measuring instrument can be affected by a number of forms of interference. These may, broadly speaking, be divided into: (a) impurities in individual components and their effects on the measuring system, and (b) injection of unwanted signals from unrelated electrical circuits and fields into the measuring system.

4.1 COMPONENT IMPURITIES

It is almost impossible to manufacture a component so that it has only one property: for example a resistance will have an associated reactance; a capacitor will have a leakage resistance; and an inductor will have a resistance per turn and an interturn capacitance. However, a particular quantity may be made dominant, and it is possible to design 'pure' components for specified limits of operation.

Under d.c. conditions of operation the quantity most likely to cause interference is the leakage resistance of insulating materials, where conduction may be either through the material or across its surface. When components have, for isolation purposes, to be mounted on insulating supports, the insulation should be of good quality and have sufficient length so that the conduction paths between the terminals of the component and between the component and earth are of negligible consequence. This is of particular importance when high voltages are in use, or when high resolution measurements are being made.

For example, if the insulation resistance between the terminals of a standard cell is 10 MΩ, its terminal voltage 1.018 50 V, and it has an internal resistance of 500 Ω, there will be a continual drain of 0.102 μA on the cell resulting in the terminal voltage being 51 μV less than the internal e.m.f. Another example of the importance of insulation resistance is in the use of instruments which have high input impedances, where for an instrument specified as having an input impedance of (say) 100 MΩ, dirty surfaces between terminals could result in this figure being reduced to 10 MΩ.

Measurements of high resistances

The measurement of high resistance is very prone to errors, because parallel leakage paths are difficult to eliminate[1], and the values obtained are affected by the magnitude of the applied voltage, the temperature of measurement, the duration of the voltage application, and the humidity. The measuring techniques used for determining high resistances may be said to fall into three categories. Let us first consider those which are an extension of the Wheatstone bridge circuit (figure 4.1).

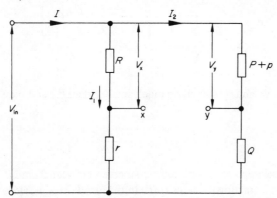

Figure 4.1. Modified Wheatstone bridge circuit

Let R be the resistance of the leakage path,

    V be the applied voltage,

    P, Q, and r the remaining arms, having magnitudes such that: $R \gg r \gg P \gg Q$.

Then with X–Y open circuit, and the bridge balanced

$$I_1 R = V_x; \quad I_2 P = V_y; \quad V_x = V_y \quad \text{and} \quad I_1 r = I_2 Q;$$

$$\therefore \qquad \frac{R}{r} = \frac{P}{Q} \quad \text{or} \quad R = \frac{rP}{Q}$$

alternatively $\qquad RQ = rP$

If the arm P is unbalanced by a small amount p,

then $\qquad V_y' = I_2 (P + p)$

and $\qquad V_{xy} = I_1 - I_2 (P + p)$

now $\qquad I = I_1 + I_2$

$$I_2 = \frac{(R + r)I}{R + P + Q + r + p},$$

and $\qquad I_1 = \frac{(P + p + Q)I}{P + Q + R + p + r}$

$$\therefore \quad V_{xy} = I \left\{ \frac{QR - (P+p)r}{P+Q+R+r+p} \right\}$$

$$= \frac{I \cdot p \cdot r}{P+Q+R+r+p} \quad \text{(Since } QR = rP\text{)}$$

now

$$I = \frac{V}{\dfrac{(R+r)(P+p+Q)}{P+Q+R+r+p}}$$

Thus

$$V_{xy} = V \left[\frac{QR - r(P+p)}{(R+r)(P+p+Q)} \right]$$

$$= \frac{Vpr}{PR + pR + pr + rQ}$$

The terms pR, pr, and rQ will all be much smaller than PR and may therefore be neglected. Giving

$$V_{xy} = \frac{V \cdot r \cdot p}{RP} \text{ V} \tag{4.1}$$

Thus the voltage applied to a detector connected between X and Y will be proportional to a fractional charge (p/P) in the arm P, and provides a method of measuring resistances up to 100 GΩ. For example, using a 1 kV supply, $r = 1$ MΩ, $P = 1$ kΩ, $Q = 1$ Ω, and $p = 10$ Ω, gives

$$V_x = \frac{1}{100} \text{ V/G } \Omega$$

The detector used in such a circuit may be a sensitive galvanometer (sensitivity 1 mm deflection per pA) or alternatively an electronic detector could be used, when the input current could be as low as 0.1 pA.

Such an instrument would have a very high input impedance (for example the input stage could contain an electrometer valve (see page 27)), and this provides another method of determining insulation resistance in which the sensitive detector is used to measure the voltage across a known resistor r which forms the low resistance part of a voltage divider consisting of r and the insulation resistance. For example, an arrangement similar to that shown in figure 4.2 from which

$$R = \left(r \frac{V_{in}}{V_r} - 1 \right) \tag{4.2}$$

Commercial instruments built using this principle have a scale calibrated in MΩ and may be used to measure resistances between 1 MΩ and 100 TΩ.

The second category of high resistance measurement methods are those which are variations of the hand generator ohmmeter, the most generally known of these being the 'Megger'[2] (manufactured by Evershed and Vignoles) which is a modified moving coil meter calibrated in MΩ and operated in conjunction with a hand or motor driven 500 V or 1000 V d.c. generator.

The third category of insulation resistance measuring instruments have evolved from techniques which use the rate of loss of charge from a capacitor. The simplest form of this type of measurement is that which involves the measurement of the time required to discharge a good quality capacitor through the high

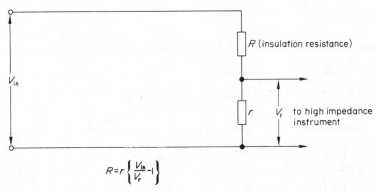

$$R = r\left\{\frac{V_{in}}{V_r} - 1\right\}$$

Figure 4.2. Potential divider method of measuring insulation resistance

resistance under test, that is assuming that the capacitor is perfect and has a capacitance C, a charge Q, and a terminal voltage V at a time t, the discharge current through the unknown resistance R is

$$i = \frac{V}{R} = \frac{dQ}{dt} = -C\frac{dV}{dt}$$

$$\therefore \quad \frac{V}{R} + C\frac{dV}{dt} = 0$$

Solving this equation gives

$$R = \frac{t}{C\log_e\left(\frac{V_0}{V_t}\right)} \tag{4.3}$$

V_0 being the capacitor terminal voltage at time = 0; and V_t the terminal voltage at time t.

It may be shown that this method of high resistance measurement is not particularly suitable for determining values of insulation resistance due to the variation of stored dielectric charge[1] with duration of electric stress. This property known as dielectric dispersion effects the apparent capacitance of the sample, whilst the dielectric dispersion itself is affected by contamination in the dielectric, for example by moisture in cellulose (paper) insulation. The dispersion of a dielectric may also be defined as the fractional change in capacitance with frequency, that is

$$\frac{(C' - C'')}{C'}$$

where C' and C'' are measured capacitances at lower and higher frequencies f' and f''. An instrument which is capable of rapidly measuring this fraction as a voltage ratio is the E.R.A. (Electrical Research Association) dispersion meter[3].

Frequency effects

The effects of frequency on the permittivity and dielectric loss of an insulating material are of consequence and figure 4.3 shows typical characteristics. These

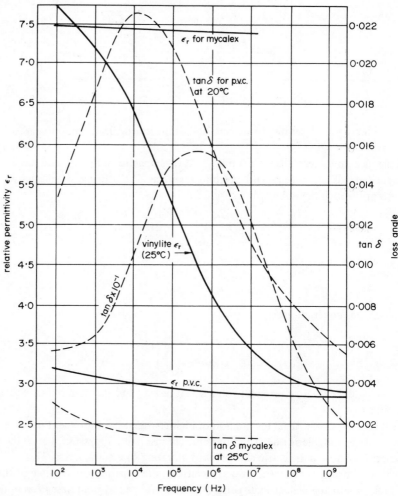

Figure 4.3. Frequency characteristics of some insulating materials (see reference 5)

variations are largely due to polarisation effects of which there are two types; dipole and interfacial. Dipole polarisation occurs in dielectrics having polar molecules, and causes the permittivity and loss angle to be appreciably affected by variations in both temperature and frequency. Interfacial polarisation (also known as dielectric absorption) occurs in composite dielectrics such as mica, but the effects of frequency on the permittivity and loss angle of this type of material are generally not as large as the effects on a dipole material

Resistor impurities

The use of resistors and resistive networks in instrumentation is extensive, and since many resistors are constructed by winding resistance wire on to bobbins, they will possess an amount of inductance. This inductive impurity of a resistor may be greatly reduced by using one of the various special noninductive forms of winding[4] (for example bifilar, woven mat, Ayrton–Perry, etc.), but a small amount of inductance will remain together with a self capacitance. The resulting

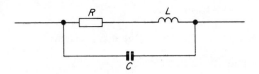

Figure 4.4. Equivalent circuit of a resistor at low and medium frequencies

impedance, which is predominantly resistive, may be represented (at medium and low frequencies) by the circuit in figure 4.4 from which

$$Z = \frac{\dfrac{1}{j\omega C} \cdot (R + j\omega L)}{R + j\omega L + \dfrac{1}{j\omega C}} \tag{4.4}$$

$$Z = \frac{R + j\omega (L - \omega^2 L^2 C - CR^2)}{1 + \omega^2 C^2 R^2 - 2\omega^2 LC + \omega^4 L^2 C^2} \tag{4.5}$$

Now for a resistor that has been constructed so that its inductance and capacitance are small, $\omega^2 LC \ll 1$ and the $\omega^4 C^2 L^2$ term may be neglected as the square of a small number, so that equation 4.5 may be simplified to

$$Z = \frac{R + j\omega [L(1 - \omega^2 LC) - CR^2]}{1 + \omega^2 C (CR^2 - 2L)} \tag{4.6}$$

Splitting Z into the real and imaginary terms gives an effective resistance

$$R_{\text{eff}} = \frac{R}{1 + \omega^2 C (CR^2 - 2L)} \tag{4.7}$$

and an effective reactance

$$X_{eff} = \frac{\omega \left[L (1 - \omega^2 LC) - CR^2 \right]}{1 + \omega^2 C (CR^2 - 2L)} \tag{4.8}$$

Now X_{eff} will be small and therefore the term $\omega^2 LC$ may be dropped from the numerator of (4.8) then

$$X_{eff} = \frac{\omega (L - CR^2)}{1 + \omega^2 C (CR^2 - 2L)} \tag{4.9}$$

The phase angle of a resistor is often of importance. Let the phase angle be ϕ, then from equations 4.7 and 4.9

$$\tan \phi = \frac{X_{eff}}{R_{eff}} = \frac{\omega (L - CR^2)}{R} = \frac{\omega (L - CR)}{R} \tag{4.10}$$

It is interesting to note that $(L/R - CR)$ is the time constant of the resistor.
From equation 4.7 it is apparent that if

$$CR^2 = 2L$$

$$R_{eff} = R \quad \text{(the d.c. resistance)}$$

Also from equation 4.9 it is desirable that

$$L = CR^2$$

so that the resistor has zero reactance and zero phase angle. These minimum impurity conditions cannot be met simultaneously, it being usual in practice to accept the very small error in resistance value that occurs when the condition giving zero phase shift is fulfilled. Then

$$R_{eff} = \frac{R}{1 - \omega^2 CL} \tag{4.11}$$

At high frequencies the resistor must be represented by an equivalent circuit of distributed components (figure 4.5). This circuit is a simplification, for a complete analysis should include the effects of interturn capacitance, and mutual couplings between turns. The solution of this type of network is performed using the techniques found in transmission line theory.

Inductor properties

The equivalent circuits of an inductor are the same as those for a resistor, except that the relative magnitudes of the components will be different. Thus at low and medium frequencies the impedance of an inductor is obtained by using the same mathematical expression as for a resistor, that is equation 4.5 which is

$$Z = \frac{R + j\omega \left[L (1 - \omega^2 LC) - CR^2 \right]}{1 + \omega^2 C^2 R^2 - 2\omega^2 LC + \omega^4 L^2 C^2}$$

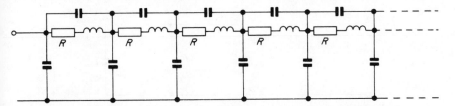

Figure 4.5. Equivalent circuit of a resistor for high frequency analysis

Now in the construction of an inductor, the resistance and capacitance will be the components that are kept to a minimum, so that if in the above expression the products of small quantities are neglected, (that is the terms $\omega^2 C^2 R^2$ and CR^2)

$$Z = \frac{R + j\omega L (1 - \omega^2 LC)}{1 - 2\omega^2 LC + \omega^4 L^2 C^2} \qquad (4.12)$$

giving an effective resistance

$$R_{\text{eff}} = \frac{R}{(1 - \omega^2 LC)^2}$$

and using the binomial expansion

$$R_{\text{eff}} \approx R (1 + 2\omega^2 LC) \qquad (4.13)$$

Now, the effective reactance

$$X_{\text{eff}} = \frac{\omega L (1 - \omega^2 LC)}{(1 - \omega^2 LC)^2}$$

which when expanded using the binomial expansion is

$$X_{\text{eff}} \approx \omega L (1 + \omega^2 LC) \qquad (4.14)$$

and the effective inductance

$$L_{\text{eff}} = L (1 + \omega^2 LC) \qquad (4.15)$$

Thus showing that both the effective resistance and inductance of an inductor increase with frequency. As with the resistor this analysis is a simplification and can only be said to apply for low and medium frequencies. At high frequencies the inductor should be treated as consisting of distributed parameters and resolved by using transmission line analysis techniques.

Capacitor properties
The effects of frequency on a capacitor are twofold. First there are the effects due to the connections to and within the capacitor, which require the inclusion in the equivalent circuit (figure 4.6) of a resistor and an inductance (R and L). The other effects of frequency on a capacitor are those effects of frequency

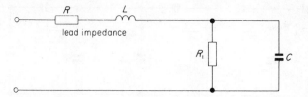

Figure 4.6. Equivalent circuit of capacitor

already referred to on the behaviour of dielectrics (see page 122). Thus the resistor R_1 representing the leakage resistance of the capacitor in the equivalent circuit will have a value which is a function of frequency. However in the following analysis it is assumed that the capacitor is to be operated at a low or medium frequency such that R_1 may be considered as having constant magnitude.

$$\text{Then } Z = R + j\omega L + \frac{\dfrac{R_1}{j\omega C}}{R_1 + \dfrac{1}{j\omega C}}$$

$$= R + \frac{R_1}{1 + \omega^2 R_1^2 C^2} + j\omega \left\{ L - \frac{CR_1^2}{1 + \omega^2 R_1^2 C^2} \right\} \tag{4.16}$$

The effective reactance $= \dfrac{1}{j\omega C_{\text{eff}}}$

$$\therefore \quad \frac{1}{\omega C_{\text{eff}}} = - \omega L + \frac{\omega C R_1^2}{1 + \omega^2 R_1^2 C^2}$$

or

$$\omega C_{\text{eff}} = \frac{1 + \omega^2 R_1^2 C^2}{\omega C R_1^2 - \omega L (1 + \omega^2 R_1^2 C^2)} \tag{4.17}$$

and providing that $\omega^2 R_1^2 C^2$ is large compared with 1

$$\omega C_{\text{eff}} = \frac{\omega C}{1 - \omega^2 CL}$$

The effective capacitance (medium frequencies) is

$$C_{\text{eff}} \approx C (1 + \omega^2 CL) \tag{4.18}$$

Note. At the frequency $\omega = 1/[LC]^{\frac{1}{2}}$ a resonant condition will be produced; it is therefore important that the inductance of leads is kept to a minimum so that the resonant frequency is high.

The effective series resistance (medium frequency) is

$$R_{\text{eff}} = R + \frac{R_1}{1 + \omega^2 R_1^2 C^2} \tag{4.19}$$

The loss angle δ is obtained from

$$\tan \delta = \frac{1}{\omega C_{eff} R_{eff}} = \frac{(\omega C R_1^2 - \omega L - \omega^3 R_1^2 C^2 L)}{R + \omega^2 R_1^2 RC + R_1}$$

$$\approx \frac{1 - \omega^2 CL}{\omega R + \dfrac{1}{\omega C R_1}} \tag{4.20}$$

At low frequencies the effects of the series inductance L and the series resistor R are negligible providing that external connections are kept as short as possible. Then from equations 4.14 and 4.16:

The effective series resistance (low frequency) is

$$R_{eff} = \frac{R_1}{1 + \omega^2 C^2 R_1^2} \tag{4.21}$$

and the effective capacitance (low frequency) is

$$C_{eff} = C + \frac{1}{\omega^2 C R_1^2} \tag{4.22}$$

As the frequency applied to a capacitor is increased the values of capacitance and $\tan \delta$ show a fall in magnitude until the effects of lead inductance become important, when an increase in loss angle and effective capacitance will occur[4-7]. Figure 4.7 shows the form of characteristic to be expected from a capacitor with a mica dielectric. Mica (a composite dielectric) exhibits a small

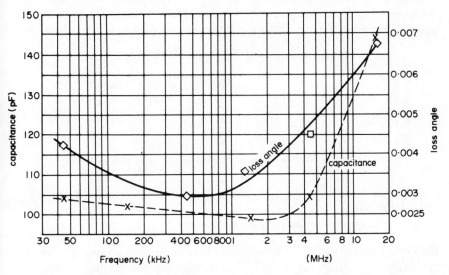

Figure 4.7. Frequency characteristics of a 100 pF mica dielectric capacitor —
measured using a 'Q' meter

amount of dielectric absorption (interfacial polarisation) and is the most suitable dielectric for use in capacitors where a high degree of stability is required.

4.2 COMPONENT IMPURITY EFFECTS ON SIGNALS

There is a tendency to think of instrument input impedances as purely resistive. From the above it is apparent that this is unlikely to be true, and the reactive component of an instrument's input impedance could introduce both a phase angle error and a frequency dependent error into the measurement. Whilst in general the phase angle error is of negligible consequence (the quantity of interest being the modulus of the measurand) the change with frequency occurring in the input impedance of an instrument cannot be neglected; since an inductive component in a measuring circuit will cause the impedance to increase with frequency and a capacitive reactance results in a decreasing impedance with frequency. Both these effects are likely to vary with the instrument's sensitivity, while another effect may result if the measuring circuit contains both inductance and capacitance in series, for then at some frequency

$$f = \frac{1}{2\pi \, [LC]^{\frac{1}{2}}}$$

there will be resonance.

Measurement of transients

Since the performance of a measuring circuit containing reactance will be affected by frequency, the effects of reactance in a measuring circuit will be of particular importance if transient phenomena are to be investigated. Consider first the behaviour of a capacitor when subjected to a sudden, or step, change in

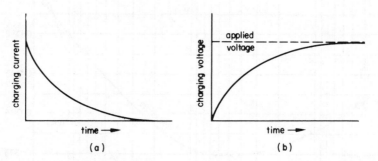

Figure 4.8. Relationships of voltage and current to time when a voltage step is applied to an uncharged capacitor

the level of applied voltage. If the capacitor initially has zero charge, then on application of a voltage current will flow into the capacitor, the magnitude of the current decreasing exponentially with time (figure 4.8a). During this charging time the voltage across the capacitor will increase until it is equal to the applied voltage (figure 4.8b). Should the charged capacitor be disconnected from the voltage source and then connected across a resistor of value R, the capacitor C will discharge with characteristics of the form shown in figure 4.9 and time constant $T = RC$ seconds, for example if $C = 1\mu F$; $R = 100\ k\Omega$ then

$$T = RC = \frac{1 \times 100\ 000}{1\ 000\ 000} = 100\ ms$$

and the time for the voltage and current change to be completed to within 1 per cent being about $5T$.

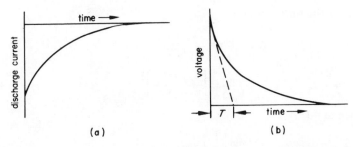

Figure 4.9. Voltage and current characteristics of a discharging capacitor

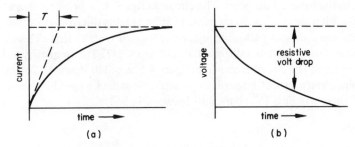

Figure 4.10. Transient response of an inductive circuit to a step function of voltage

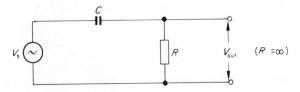

Figure 4.11. 'High pass' filter circuit

Figure 4.10 shows the response of an inductive circuit to a step change in voltage. It being noted that for the inductive case the time constant of the circuit is $T = L/R$.

Thus it will be apparent that any system for recording transients must contain a minimum of reactive impedance or distortion of the measurand will result, for the time constant of a circuit is directly related to the frequency response or bandwith of a circuit.

Consider the circuit in figure 4.11 which has open circuited output terminals (that is $R_L = \infty$).

Then

$$\frac{V_{out}}{V_s} = \frac{R}{R + \dfrac{1}{j\omega C}} = \frac{j\omega CR}{1 + j\omega CR}$$

now $CR = T$ the time constant

$$\therefore \quad \frac{V_{out}}{V_s} = \frac{j\omega T}{1 + j\omega T}$$

or
$$\left|\frac{V_{out}}{V_s}\right| = \left[\frac{\omega^2 T^2}{1 + \omega^2 T^2}\right]^{\frac{1}{2}} \tag{4.23}$$

now when $\omega = 1/T$

$$\left|\frac{V_{out}}{V_s}\right| = 0.707 \tag{4.24}$$

and this defines the -3 dB point. The circuit in figure 4.11 being termed a 'high pass' filter meaning a circuit capable of preventing the passage of low frequency and d.c. whilst allowing high frequency currents to flow. This circuit may also be used as a differentiator, if $\omega \ll 1/T$. The frequency output characteristic of the circuit is shown in figure 4.12a and in figure 4.12b, a sketch of the phase angle against frequency characteristic is drawn and shows that at $\omega = 1/T$ there will be a $45°$ phase shift between V_s and V_{out}, that is,

$$\text{phase angle } \phi = \tan^{-1}\frac{1}{\omega CR}$$

$$= \tan^{-1}\frac{1}{\omega T} \tag{4.25}$$

and if
$$\omega = 1/T \; \phi = \tan^{-1} 1 = 45°.$$

Consider now the circuit in figure 4.13. Inspection of this circuit gives

$$\frac{V_{out}}{V_s} = \frac{\dfrac{1}{j\omega C'}}{R' + \dfrac{1}{j\omega C'}} = \frac{1}{1 + j\omega C'R'} \tag{4.26}$$

If the time constant of the circuit is T', then

$$\frac{V_{\text{out}}}{V_{\text{s}}} = \frac{1}{1 + j\omega T'}$$

and when $\omega = \dfrac{1}{T'}$ $\quad \left|\dfrac{V_{\text{out}}}{V_{\text{s}}}\right| = 0.707$

while at $\omega = 0$ $\quad \left|\dfrac{V_{\text{out}}}{V_{\text{s}}}\right| = 1$

and at $\omega = \infty$ $\quad \left|\dfrac{V_{\text{out}}}{V_{\text{s}}}\right| = 0$

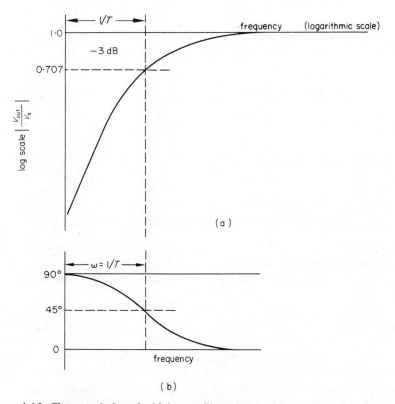

Figure 4.12. Characteristics of a high pass filter circuit. (a) Response—frequency characteristic; (b) phase angle—frequency characteristic

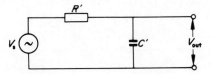

Figure 4.13. 'Low pass' filter circuit

This circuit is known as a 'low pass' filter or alternatively as an integrator (if $\omega \gg 1/T$) (see page 174), and will have characteristics of the type shown in figure 4.14.

It should be appreciated that circuits of both the above types may occur at a number of places in an instrumentation system. For example the circuit in

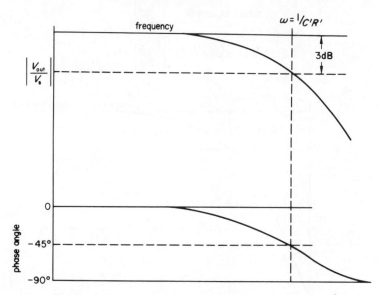

Figure 4.14. Characteristics of low pass filter circuit

figure 4.13 is the equivalent circuit of a source of resistance R feeding into a coaxial cable, or c.r.o. The circuit of figure 4.11 is that which occurs at the input to an a.c. amplifier, or at the output of some amplifiers when supplying a resistive load.

Combinations of resistance and inductance may also be shown to behave in a somewhat similar manner. Figure 4.15a being a 'high pass' circuit and figure 4.15b a 'low pass' circuit, for at low frequencies the inductor will tend to have zero impedance, whilst at high frequencies its impedance will tend to infinity.

These resistance — reactance combinations will affect the shape of a pulse or series of pulses as they pass through the circuit. For example, applying a

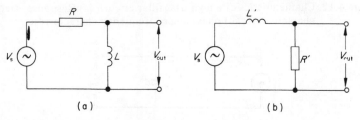

(a)   (b)

Figure 4.15. Inductive (a) high pass and (b) low pass circuits

rectangular pulse to a low pass circuit will result in distortion as illustrated in figure 4.16; the risetime, t_r, of the distorted wave being defined as the time interval from 10 per cent to 90 per cent of the maximum value and equal to $2.2T$ s, where $T = CR$ (or L/R) the time constant of the circuit (see also page 129). Applying a pulse to a high pass circuit results in a distortion where the low frequency cut off causes the top of the rectangular pulse to 'sag' as in figure 4.17, the fractional sag being approximately t_d/T for a single waveform, where t_d is the pulse duration.

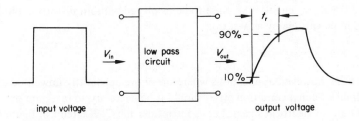

Figure 4.16. Distortion of a rectangular pulse through a low pass circuit

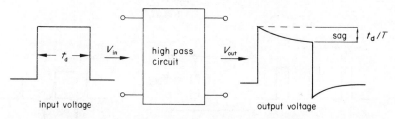

Figure 4.17. Distortion of a rectangular pulse through a high pass circuit

If both high pass and low pass circuits (formed from resistance and capacitance) are present, the output waveform will have a shape similar to that shown in figure 4.18a. Should the connections to the combination of circuits possess a significant amount of inductance then a wave shape of the form shown in figure 4.18b will result. To compensate for these undesirable distortions of pulse shape, c.r.o. and other leads may be fitted with probes that contain a circuit of the type shown in figure 4.19, and adjustment of C_1 (say) such

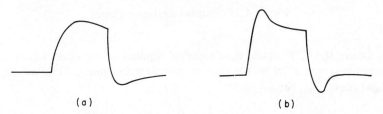

(a)   (b)

Figure 4.18. Distortion of a rectangular pulse by both high and low pass circuits
(a) and (b) with the addition of lead inductance

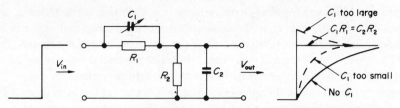

Figure 4.19. Effect of the magnitude of C_1 on the output waveform

that $C_1R_1 = C_2R_2$ enables the pulse to be displayed without distortion but reduced in magnitude by a factor of

$$\frac{R_2}{R_1 + R_2}$$

There are, however, applications where the above apparently unwanted effects are desirable. Such a condition arises should it be necessary to integrate or differentiate an electrical signal. Let us reconsider the C–R circuit in figure 4.11 for which

$$\frac{V_{\text{out}}}{V_{\text{in}}} = \frac{j\omega CR}{1 + j\omega CR}$$

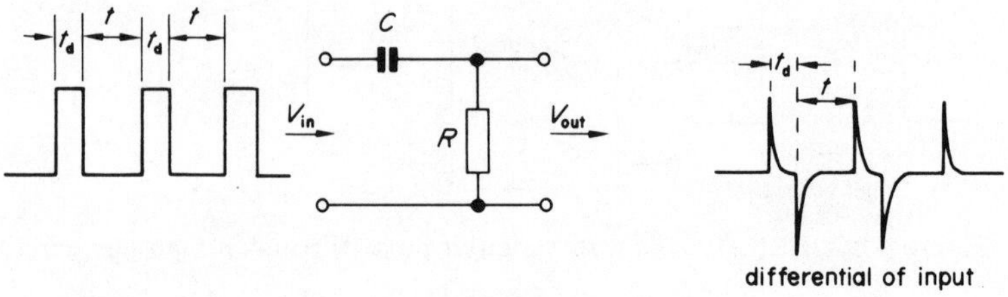

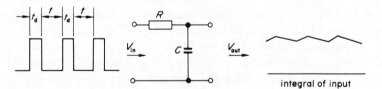

Figure 4.20. Simple differentiation circuit

Figure 4.21. Simple integration circuit

and assume the V_{in} is a pulse train made up of pulses having a duration of t_d and separation t; then if $CR \ll t$ the output waveform in figure 4.20 will result, the magnitude of V_{out} being

$$V_{\text{out}} = \left| \frac{R}{R + \dfrac{1}{j\omega C}} \right| V_{\text{in}}$$

This output waveform is the differential of the input waveform and has a number of applications in instrumentation. For satisfactory differentiation the CR product must be equal to or less than $t/100$.

The potential difference across a capacitor is a measure of the integral with respect to time of the current that has flowed into it, thus the $R-C$ circuit of figure 4.21 may be used as an integrator providing that $CR \gg t$. For example a train of rectangular pulses would be integrated as shown in figure 4.21.

4.3 PROBES

The purpose of a probe is to connect an instrument, such as an oscilloscope or electronic voltmeter, to a test circuit in a manner such that the presence of the monitoring instrument does not affect the circuit under investigation. In many applications the probe consists of a coaxial lead terminated with insulated 'prods'. A coaxial lead being used to prevent 'pickup' distorting the signal that is being measured. The instrument has a high impedance so that any loading effects of the instrument on the circuit under test are negligible. This latter condition may not be satisfactorily fulfilled if the test circuit has either a very high impedance, or is

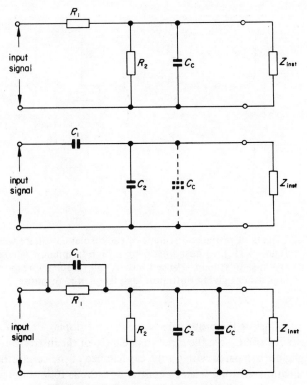

Figure 4.22. Input 'attenuator' probes

operating at a very high frequency, when the capacitance of the coaxial connection will introduce a low impedance across the instrument's input impedance. A probe containing either an input 'attenuator' or a cathode follower must then be used to increase the impedance that the test circuit 'sees'. The input 'attenuator' type of probe will introduce a reduction in signal amplitude. The type illustrated on page 134

gives a reduction in signal level of $\dfrac{R_2}{R_1 + R_2}$; and this is typically $\dfrac{1}{10}$.

The probe may be a resistance, capacitance, or combined resistance − capacitance divider[8] (figure 4.22). The latter form will have an adjustable component so that uniformity of frequency response may be obtained. Figure 4.23 illustrates

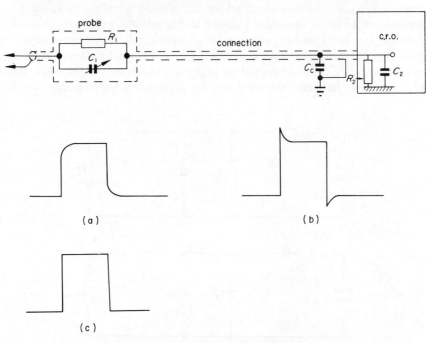

Figure 4.23. Effects of probe adjustment of the displayed square wave (see also figure 4.19.) (a) misadjustment − high frequency attenuation (b) misadjustment − accentuated high frequency response (c) correct adjustment giving optimum high frequency response

the effects of adjusting the probe capacitance on the display of a square wave. In many cases R_2 and C_2 are the input impedances of the instrument, and C_2 may be negligible compared with C_c the capacitance of the connecting cable.

The attenuation resulting from an emitter or cathode follower probe will be small, as such a device has approximately unity gain but is capable of having an input impedance many times greater than its output impedance, thus increasing the impedance the test circuit 'sees'.

For measuring a.c. current waveforms on an oscilloscope a current probe[8] may be used. This is a form of current transformer, the core of which may be opened to allow the current carrying conductor to be inserted (figure 4.24a), giving a probe sensitivity of 1 mA/mV over a bandwidth from 1 kHz to 200 MHz. A modification to this circuit, figure 4.24b, is to incorporate a Hall effect device

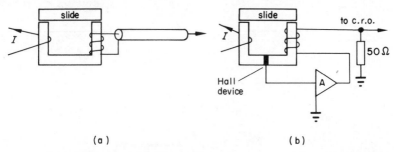

(a)   (b)

Figure 4.24. (a) a.c. and (b) d.c. oscilloscope current probes. (Tektronix)

(see page 39) in the *CT* core so that the bandwidth may be extended down to d.c., the upper frequency limit being 30 MHz.

A further type of probe is the detector or demodulator type (see page 3). This essentially consists of some form of diode circuit so that either a high frequency bias may be removed from an input signal or the peak value of the signal determined.

4.4 ELECTROSTATIC INTERFERENCE

When a conductor A is maintained at a potential relative to zero (earth) potential, it will be enveloped in an electrostatic field (see figure 4.25a). If another conductor B, insulated from earth is introduced into this electrostatic field it will attain a potential dependant on its size and its position in the field (see figure 4.25b), and should the potential of conductor A be alternating then the conductor B will 'pick up' an alternating voltage. This linking by means of an electrostatic field is termed capacitive coupling and may cause considerable interference in measurement circuits. To shield the conductor B from the electrostatic field of conductor A requires the insertion of a conducting sheet, held at zero potential, between the two conductors (see figure 4.25c).

In a practical measuring system this principle must be extended so that both the connections between, and the component parts of, the system are contained within the shield (see figure 4.26). Such an arrangement is an ideal, being a simplification of the conditions that are probable in practice where, for example, both the source and the measuring instrument may have one terminal permanently connected to the power supply earth, in addition to this any practical screen

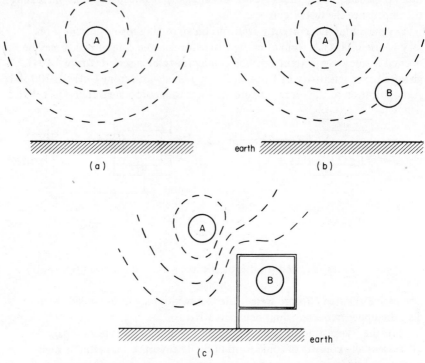

(a)   earth   (b)

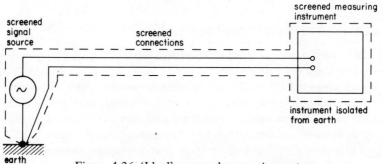

(c)

Figure 4.25. Conductor and shield in an electrostatic field

screened measuring
instrument

screened
signal   screened
source   connections

instrument isolated
from earth

earth

Figure 4.26. 'Ideal' screened measuring system

will have a resistance and an inductance, and earth currents flowing through it
will cause a potential gradient along the length of the screen. Since there will be a
capacitive coupling between the screen and the screened conductor or components
a certain amount of interference may result. It is therefore necessary to consider
the conditions that are likely to occur in practice, although it should be realised
there are an almost infinite number of combinations of connections, components
and arrangements. It is therefore only possible to indicate general principles.

Instrument earth connections

It has been stated that a measuring instrument may in practice have an earthing arrangement different from that of the ideal shown in figure 4.26 and since the measuring instrument is likely to be the most expensive component of a measuring system, it is sensible to arrange the screening of the system based on the measuring instrument's earth. This 'earth' will be combined with the input terminals in one of the following ways:

(a) Instruments with two input terminals, one of which is connected to the power line earth. This type of arrangement, which is used on many popular instruments (for example many general purpose oscilloscopes and valve voltmeters), means that only signals from sources *isolated* from earth (figure 4.27a) may be measured without introducing errors due to interference voltages. The alternative to this is that the instrument measures the signal plus any interference e.m.fs, for example, those due to current circulating around the 'earth' loop formed between the earth connections and the screen (figure 4.27b). However these interference voltages may be negligible compared with the magnitude of the measurand.

(b) Instruments with an isolated input, there being an additional 'earth' input terminal which is connected, via the instrument chassis, to the supply earth. The terminal markings on this type of instrument are usually Hi (high), Lo (low), and earth (or ground) where the Lo terminal may be a *virtual earth* or linked to earth via a resistor (see page 173). The instrument may be fitted with an input lead which could either be a double coaxial lead (or a screened twin conductor), figure 4.27c and d being typical connections. Should the signal source have an output via a coaxial connection, the three conductor input lead of the instrument should be connected to it as shown in figure 4.27e or f.

(c) The fulfilment of the ideal, where the only earth connection is on one side of the signal source, may only be obtained when instruments are used that have no connections with external supplies, for example, battery powered instruments, or are fitted with special guard circuits which isolate the instrument and its screen from the power supply earth. That is power is fed into the instrument via a screened isolating transformer (figure 4.28). This sort of screening technique is expensive and is normally resorted to in expensive instruments, for example in some sensitive digital voltmeters.

It will be apparent from the above that there are three common forms of electrostatically screened connection, namely the coaxial lead, the double coaxial lead, and the twin conductor with screen. The last of these is sometimes termed microphone cable and is only suitable for d.c. and low frequency applications; while the coaxial (single and double) are suitable for use from d.c. to

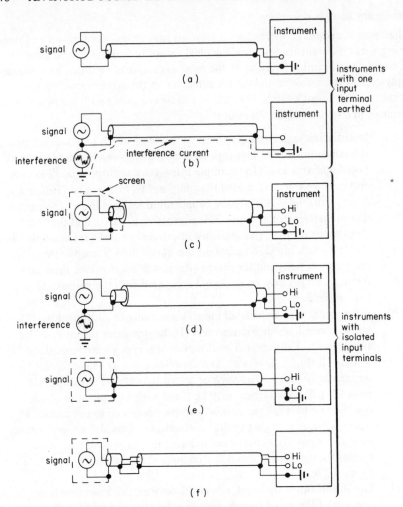

Figure 4.27. Practical screen connections for measurement systems

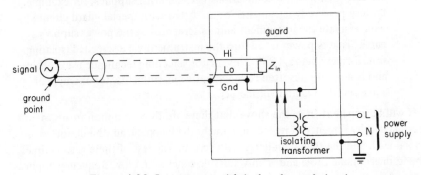

Figure 4.28. Instrument with isolated guard circuit

gigahertz, although at high frequencies care may be necessary to match the impedance of the cable to that of connectors to avoid reflections and distortion of the signal.

Signal conditioning circuits may require external power, for example when operational amplifiers are involved (see page 171), it may also be desirable to screen and isolate them from the power supply. A few remarks relating to the use of screened isolating transformers are therefore relevant. A screened isolating transformer will normally have either a single or a double screen, when constructed with the former it is usual to connect the screen either into the screen circuit[9] (figure 4.29a), or to the midpoint of the secondary winding (figure 4.29b).

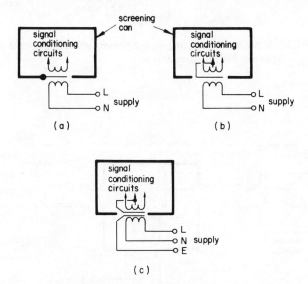

Figure 4.29. Isolating transformer screen connections

Neither of these arrangements removes all the interference, and while a single screen is satisfactory for some applications the best results are obtained when a double screen is used and connected as shown in figure 4.29c.

Amplifier screening

The connection of guard or screen circuits to and around amplifiers are of considerable importance for if interference voltages are applied to their input the magnitude of these unwanted signals will be increased along with the signal from the measurand. Consider first the use of a single ended amplifier as part of a measurement system comprising of sensor, signal conditioning, and display or record instrument. If the transducer (see page 230) has a single coaxial lead output then the connection arrangement should be as in figure 4.30, but when the transducer has a 'twin plus earth' output the connections used are as shown in figure 4.31. Whilst both these arrangements reduce the interference from that

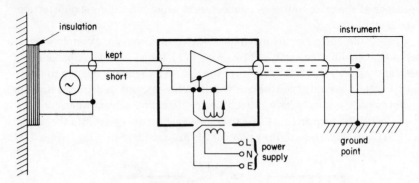

Figure 4.30. A practical shielding arrrangement when the source has a single coaxial output connection

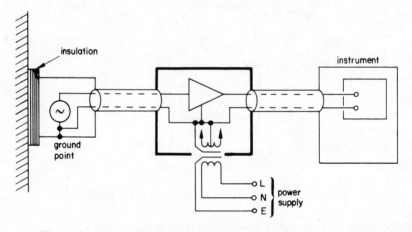

Figure 4.31. A superior arrangement to that shown in figure 4.30

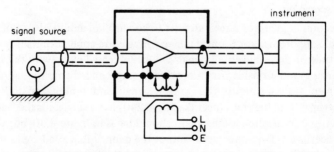

Figure 4.32. Application of a segmented amplifier shield

which would exist if no shielding were employed; a greater reduction still may be obtained by using segmented shielding, such as that shown in figure 4.32 where the amplifier shield is held at the same potential as the zero (low) potential input lead of the amplifier. It must, however, be appreciated that should it be desired to use a display instrument which has an earthed terminal when a single ended amplifier is in use, then the transducer or signal source must be isolated from earth. If an amplifier with a differential input is available, this may be used to ovecome the problem, but a more common application of the differential amplifier is in situations where both leads from the signal source have a potential to earth, or 'common mode' voltage, for example the output leads from an unbalanced Wheatstone bridge (figure 4.33)[11].

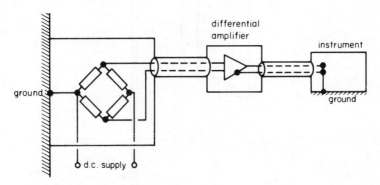

Figure 4.33. Use of a differential amplifier, when both source and instrument have an earth point

A method used, in some investigations, to screen the apparatus from electrostatic interference is a 'Faraday Cage' — this being a room or enclosure constructed from wire mesh (preferably aluminium or copper). Since no electrostatic field from an external source can exist within a metal container, the apparatus within the cage will only be subject to electrostatic fields produced within the cage, providing power supplies are conducted into it via screened isolating transformers. When using this technique to shield from electrostatic interference moderately sized holes may be cut in the enclosing screen without impairing the effectiveness of the screen.

One further application of electrostatic screening that should be mentioned is the use of guard rings in the control of the magnitude of a capacitance. Figure 4.34 illustrates that without the use of an earthed guard ring the value of capacitance between the plates A and B could only be calculated approximately. Because of the unknown magnitude of the fringing effects, however, when an earthed guard ring is added the field between A and B is more defined and agreement between calculated and measured values of capacitance will be much closer. Possibly of greater value will be the fact that external fields and objects will have very much less effect on the magnitude of the guarded capacitor than these

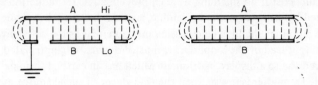

Figure 4.34. Circular parallel plate capacitor, with and without a guard ring

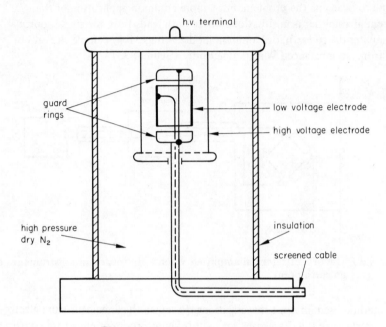

Figure 4.35. High voltage capacitor

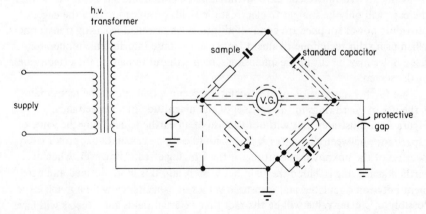

Figure 4.36. Shielding arrangement of a high voltage Schering bridge

effects on an unguarded capacitor. Figure 4.35 is a diagrammatic section of a high voltage capacitor, fitted with guard rings, and filled with gas at a pressure above atmospheric. The connection of such a capacitor into a screened Schering bridge circuit is given in figure 4.36 (see also page 93).

4.5 ELECTROMAGNETIC INTERFERENCE

Any current carrying conductor is surrounded by a magnetic field, the strength of this field depending on the magnitude of the current, and should it be alternating, voltages will be induced in any conducting material that is positioned within this magnetic field. Thus any signal carrying connection passing near a current carrying conductor will be magnetically coupled to it, and an interference voltage will be induced in the signal carrying conductor. One of three forms of magnetic shielding may be used to reduce electromagnetic interference, these being:

(a) *Conducting shield.* The positioning of a thick conducting sheet around the component or lead to be screened; then magnetic flux entering the conducting sheet induces currents which themselves produce a magnetic flux to oppose the applied field.

(b) *Magnetic shield.* The positioning of a high permeability path around the component to be screened so that the magnetic flux of the interference field is diverted into the shield and away from the component. (Note. The thickness of the shield required in both (a) and (b) will be a function of frequency, and field strength.)

(c) *Lead twisting.* By arranging equipment so that the coupling between the magnetic field source and the signal carrying connections is either cancelled out, or kept to a minimum. The former requirement may be met by ensuring that the two leads connecting a source to an instrument follow an identical path, and such an arrangement is greatly assisted by twisting the two leads together. The latter requirement can only be fulfilled by ensuring that the signal carrying conductors are remote from any source of magnetic flux. In practice the most likely sources of electromagnetic interference are 50 Hz transformers and inductors, and all signal carrying conductors should be kept well clear of such items. If two, or more, inductors have to be used in a circuit they should be arranged for minimum coupling, that is their axes should be arranged at 90° to each other.

To eliminate the effects of r.f. electromagnetic interference it may be necessary to place equipment subject to r.f. interference in a screened room or enclosure. The shield material used for such a construction is usually copper sheet, and since the prevention of electromagnetic interference is only possible if the screen is complete, all joints must be electrically continuous and of low resistance, that is there must not be any open circuits at doors, etc. Power supplied into such an enclosure should be filtered and screened (see pages 130 and 141), the filter having a shunt capacitor (low impedance at high frequency) followed by a series inductor (high impedance at high frequency).

Electrostatically screened conductors[10] (for example coaxial connections) afford a certain amount of protection from electromagnetic interference, and providing the magnetic field strength is not too large, satisfactory shielding from electromagnetic fields may be obtained by using screening intended for shielding from electrostatic fields. Similarly a shield designed for shielding from an electro-magnetic interference will afford an amount of electrostatic screening.

4.6 MULTIPLE EARTHS AND EARTH LOOPS

Ideally all the earth connections in a system are at the same (zero) potential but this can only be true if the earth path has zero resistance and inductance. This is not possible in a practical case, and any circuits which utilise the earth connec-tions as a current return path will cause a voltage gradient to exist along the earth path. Two effects result from this: first, capacitive coupling between the earth path and the signal conductor may result in electrostatic interference being added to the signal; and secondly, if the return current is added to earth currents from shields and other circuits, these will appear to the measuring circuit as an increase in signal level.

To reduce this type of interference it is vital to avoid using the earth path as a signal return path and to make sure that a circuit is only earthed at one point. For if more than one point is used and currents are flowing in the earth path, a current i will circulate through the signal circuit (see figure 4.37), the magnitude

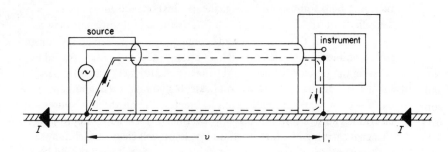

Figure 4.37. Earth loop and circulating current

of i depending on v the volt drop between earth points and the lead resistances[11,12]. This circulatory path is termed an earth loop. If, however, the instrument in use is of the guarded type (or battery operated) it is possible for v to have an appreci-able magnitude and yet have negligible effect on the displayed reading. This property in an instrument's specification is termed the *common mode rejection* (c.m.r.) and is usually expressed as the ratio between the common mode voltage, e_{cm} (see figure 4.38) and the error in the voltmeter reading E_e, due to the

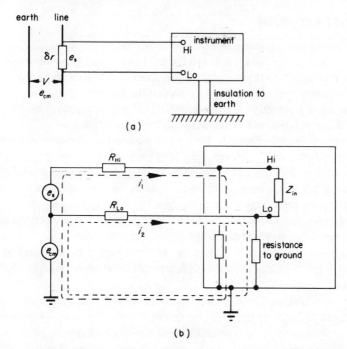

Figure 4.38. Common mode voltage effects. (a) Common mode voltage situation; (b) equivalent circuit of (a)

presence of this common mode voltage, that is

$$\text{c.m.r.} = \frac{e_{cm}}{E_e} \tag{4.27}$$

Now the common mode voltage may be a.c. or d.c., the impedances in the high and low sides of the circuit, and of the two sides to earth may contain reactive components; and all these factors will affect the magnitude of the error voltage between the Hi and Lo terminals due to the common mode voltage. It is therefore customary to quote in a specification values of c.m.r. ratio for d.c. and for particular frequencies, that at the power line frequency usually being of greatest interest.

Since the c.m.r. ratio may be large, and the attenuation effect of the instrument guard circuitry on the common mode voltage is of interest, the c.m.r. ratio is normally given in decibels (dB), that is

$$\text{d.c. c.m.r. ratio} = 20 \log_{10} \left\{ \frac{e_{cm}}{E_e} \right\} \text{dB} \tag{4.28}$$

$$\text{and} \quad \text{a.c. c.m.r. ratio} = 20 \log_{10} \left\{ \frac{e_{cm} \text{ (Peak)}}{E_e \text{ (Peak)}} \right\} \text{dB at } f \text{ Hz} \tag{4.29}$$

4.7 DRIFT AND NOISE

When a measurement system includes amplification, either voltage or current, it may be necessary to offset from zero the level about which the input signal may vary, so that with zero input signal the amplifier output is zero. This offset be it voltage or current will invariably be temperature dependent, resulting in a temperature dependent drift in the amplifier output. There may also be an output drift which is independent of temperature but time dependent.

A common source of electrical noise in a measurement system results from poor connections, which, for example, may be due to bad soldering, or dirt on switch contacts. It is therefore vital when dealing with small signal levels that great care is taken in forming joints and contacts. Voltage and current noise also result from the energy fluctuations generated by the random movement of free electrons within a conductor (Brownian movement). The spectrum of this noise covers all frequencies and is the limiting level to which noise may be reduced in a sensitive measuring system. Its effect may only be limited by making the system bandwidth as narrow as is acceptable in relation to the bandwidth of the signal.

When several signal carrying circuits are routed via a multiconductor cable, interference or crosstalk between circuits may result. This form of interference may be overcome by using screened connections for each separate circuit, this being a fairly expensive solution. However, as a compromise, it may be possible to select either the most sensitive or the most troublesome circuit and screen it separately from the remainder and reduce the crosstalk to an acceptable level.

Finally the interference effects of voltage and current noise, thermal drift, time dependent drift, thermal e.m.fs, etc., will all be dependent on the magnitude of source, feedback, and input impedances, and consequently a measurement system incorporating a signal source which has a high impedance (necessitating an instrument with a high input impedance) will be the most prone to interference effects; and in many cases these effects may only be reduced by the inclusion of an emitter follower adjacent to the signal source.

REFERENCES

1   W. P. Baker. *Electrical Insulation Measurements,* Newnes, London (1965)
2   E. W. Golding and F. C. Widdis. *Electrical Measurements and Measuring Instruments,* Pitman, London (1963)
3   G. Mole. 'Improved methods of test for the insulation of electrical equipment' *Proc. I.E.E.* **100** (IIA) p. 276 (1953)
4   F. E. Terman and J. M. Pettit. *Electronic Measurements,* McGraw-Hill, New York (1952)
5   A. R. Von Hippel (Ed.) *Dielectric Materials and Applications,* M.I.T. Press, Cambridge, Mas. (1954)
6   J. V. L. Parry. *E.R.A. Technical Reports* L/T 275 and 325

7   H. Fröhlich. *Theory of Dielectrics* (2nd edn), Clarendon Press, Oxford (1958)

8   Cal Hongel. *Plug on Versatility.* Tektronix Service Scope No. 51 (Aug. 1968)

9   R. Morrison. *Grounding and Shielding Techniques in Instrumentation,* Wiley, New York (1967)

10   B. A. Gregory. 'An Experiment in Shielding and Screening'. *I.J.E.E.E.* 9, 31–35 (Dec. 1971)

11   'Floating Measurements and Guarding', *Application Note 123,* Hewlett Packard Co. (1970)

12   C. E. Emgle. 'Techniques to Analyse and Optimize Noise Rejection Ratio of Low Level Differential Data System'. Dana Labs Inc. *Technical paper No. 521* (1965)

5

Signal Conditioning

In many instances the values of electrical quantities (voltage, current and power) are too large or too small to be connected directly to the available instrument. It therefore becomes necessary to suitably reduce, or amplify, the magnitude of the measurand so that it has a value compatible with the measuring instrument to be used. In addition to these requirements, the effects of the instrument's impedance must always be considered for this may effect the value indicated for the measurand.

5.1 IMPEDANCE EFFECTS

The magnitude of the impedance that exists between the input terminals of an instrument is extremely important. Upon this impedance will depend the disturbance that is created when the instrument is inserted into the circuit for measurement purposes. This disturbance should be minimal, that is to say, an ammeter should have as low an impedance as possible, while for a voltmeter the requirement is a high impedance.

In practice it is the relative magnitudes of the instrument's impedance compared to that of the circuit which is important. For example, a 100 Ω, 1 mA f.s.d. instrument, when connected in series with a 100 Ω load, will halve the current in the circuit; but if the same instrument is connected in series with a 100 kΩ load its effect would be negligible (see page 2). Similarly, it is just not practicable to measure the output voltage of a source which has a high internal resistance (for example 100 kΩ) with a 50 Ω/V voltmeter; but instruments with this value of input impedance are extensively used when measuring voltages associated with supply frequency power circuits, where the source impedance is low.

Impedance matching

In some instrumentation systems a factor of importance is the conveyance of the maximum amount of power contained in the output signal from a sensor into a display or recording instrument, for example to obtain the maximum deflection of a u.v. recorder galvanometer. Such a requirement necessitates the matching of the output impedance of the sensor to the input impedance of the measuring circuits. Consider a signal source, such as that shown in figure 5.1

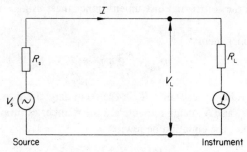

Figure 5.1. Source and instrument impedance mismatch

let V_s be the source voltage
R_s be the source resistance
R_L be the load resistance

Then
$$I = \frac{V_s}{R_s + R_L}$$

and the power transferred to the load

$$= I^2 R_L$$

$$= \frac{V_s^2 \cdot R_L}{R_s^2 + 2R_s R_L + R_L^2}$$

$$= \frac{V_s^2}{\dfrac{R_s^2}{R_L} + 2R_s + R_L} \tag{5.1}$$

To find the value of R_L that makes the load power a maximum, the denominator of equation 5.1 must be a minimum. This will occur when

$$\frac{d\left\{\dfrac{R_s^2}{R_L} + 2R_s + R_L\right\}}{dR_L} = 0$$

or
$$\frac{-R_s^2}{R_L^2} + 1 = 0$$

or
$$R_s = R_L \tag{5.2}$$

In practice this desirable state may not exist. As an example consider the circuit in figure 5.1, where the source (V_s) has an impedance (R_s) of 600 Ω, and the instrument to which it is connected has an impedance of 1000 Ω; under these conditions the power transmitted to the load is 93 per cent of the maximum power that could have been transmitted, and for the d.c. case the optimum may only be attained by changing to an instrument that has an input impedance of 600 Ω. If the signal from the source is purely a.c. (no d.c. component) it is possible to match the source and instrument impedances using a matching transformer.

From transformer theory

$$R_{\text{eqs}} = \frac{N_{\text{p}}^2}{N_{\text{s}}^2} \times R_{\text{s}}$$

where R_{eqs} is the value of resistance R_{s} in the secondary circuit referred to the primary circuit; N_{p} and N_{s} being respectively the number of primary and secondary turns. Let the equation be rewritten

$$\frac{R_1}{R_2} = \left(\frac{N_1}{N_2}\right)^2$$

where R_1 is the resistance of the circuit on the side of the transformer which has N_1 turns; R_2 and N_2 being the resistance and turns on the other side of the transformer. Then if this transformer is added to the circuit of figure 5.1 it may

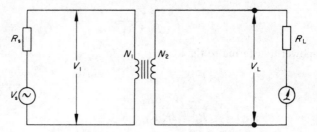

Figure 5.2. Use of a matching transformer

be redrawn as in figure 5.2, and for matching of the source and instrument, the transformer should have a turns ratio of

$$\frac{N_2}{N_1} = \left(\frac{R_2}{R_1}\right)^{\frac{1}{2}} = \left(\frac{R_{\text{L}}}{R_{\text{s}}}\right)^{\frac{1}{2}} \tag{5.3}$$

which for the above example requires

$$\frac{N_2}{N_1} = \left(\frac{1000}{600}\right)^{\frac{1}{2}} = 1.291$$

that is, the matching transformer would have to have a turns ratio of 1.291 giving $V_{\text{L}} = 1.291\ V_1$ and power transferred to the instrument

$$= \frac{V_{\text{L}}^2}{R_{\text{L}}} = \frac{(1.291)^2\ V_1^2}{1000}$$

$$= 0.001\ 666 V_1^2\ \text{W}$$

Now $V_1 = \dfrac{V_s}{2}$ (for maximum power transfer)

∴ power to the instrument

$$= 0.001\ 666\ \frac{V_s^2}{4} = 0.416\ V_s^2\ \text{mW}$$

The power available from the source

$$= \frac{V_s^2}{2R_s} = \frac{V_s^2}{2 \times 600} = 0.833\ V_s^2\ \text{mW}$$

The above shows that the maximum possible (that is 50 per cent) of the power available may be transferred to the instrument using a matching transformer. In practice, there are limitations to this principle. First, the practical transformer has losses and these must be supplied from the signal source; secondly, matching transformers are expensive and the increase in sensitivity obtained by matching may not be justifiable; thirdly, a matching transformer can only be used if the signal is alternating since d.c. signals will not pass through the transformer.

5.2 VOLTAGE SCALING

Voltage dividers are an important aspect of scaling and have been mentioned in connection with range extension of voltmeters and for use with, and within, potentiometers and multimeters. Voltage dividing resistors are sometimes confusingly termed potentiometers as an abbreviation of the term potential or potentiometric divider.

5.2.1 Resistance Dividers

The resistance chain is the simplest form of divider, the voltage division being due to the total voltage drop across the chain in comparison with the volt drop across the end unit. The chain may be connected in series with a pointer instrument (see figure 1.5 page 10) when the resistance of the measuring instrument forms part of the chain, or the measuring instrument is connected across the low voltage end of the divider. If the latter method of operation is used it is extremely important that the input impedance of the measuring instrument should be very much greater than the chain resistor across which it is connected, or a large error in the measurement will result. Consequently, this form of divider should only be used with potentiometers or electronic voltmeters that have a high input impedance. Such dividers are of course suitable for d.c. use and may be used on a.c. providing they are pure resistance (see page 123), that is the impedance remains constant irrespective of frequency. The range of d.c. voltages over which this type of divider is employed is limited by power dissipation and earth leakage path resistance.

The principle of the Kelvin–Varley divider has been outlined in chapter 3 in connection with its use in potentiometers. It is an inherently more precise method of voltage division than the simple resistance chain and finds extensive application in the standards laboratory. Considering the Kelvin—Varley divider in detail it is seen to consist of several decades of resistors interconnected as in figure 5.3. Each voltage division decade is made up of eleven equal resistors, successive

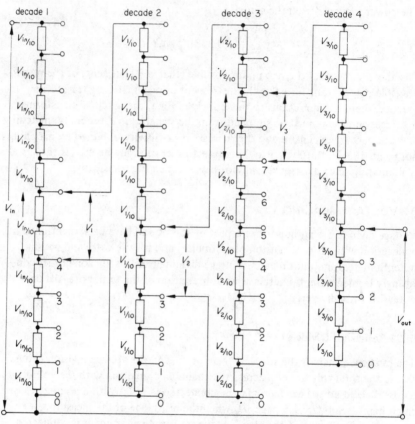

Figure 5.3. Kelvin—Varley divider set to 0.4374

division decades having a total resistance equal to twice the value of a unit resistor in the previous decade. For example, if in a four decade divider the first decade is constructed using 11 x 10 kΩ resistors, the second decade will have 11 x 2 kΩ resistors, the next decade 11 x 400 Ω resistors and the fourth decade or final division will have 10 x 80 Ω resistors. The use of 11 resistors to obtain a decade voltage division enables the Kelvin—Varley divider to have a constant input impedance (strictly with the output open circuited) irrespective of which switch positions are connected on the various decades. For example decade 3 will have a constant impedance of

$$9 \times 400 + \frac{2 \times 400 \times 10 \times 80}{800 + 800} = 2 \text{ k}\Omega$$

decade 2 an impedance of

$$9 \times 2000 + \frac{2 \times 2000 \times 4000}{80\,000} = 20 \text{ k}\Omega$$

and the input terminal impedance will be 100 kΩ irrespective of the decade switch positions (providing that the output current is small).

Another advantage of the Kelvin–Varley divider is the reduction of switch contact resistance effects due to the current sharing within the divider[1]. The drawbacks of the device are calibration (which can be simplified) and temperature distribution in the resistance chains (since all the resistors of a chain do not carry the same current they will not all operate at the same temperature). This latter effect can be reduced to negligible proportions by using resistors of low temperature coefficient, resulting in dividers with a total uncertainty of ± 0.1 p.p.m.

5.2.2 Capacitive Divider

The voltage dividers so far described have been mainly applicable for d.c. and low frequency a.c. The capacitive divider is basically unsuitable for d.c. use, since voltage division would then rely on leakage current, but it may be used on a.c. from power frequencies to MHz. It consists of a chain (usually 2 in series) of capacitors, the voltage division being inversely proportional to the magnitudes

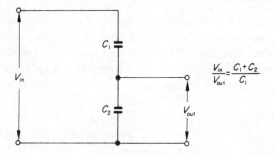

Figure 5.4. Capacitive divider

of capacitance. Figure 5.4 shows a capacitor divider consisting of a 100 pF capacitor in series with a 0.1 μF capacitor, the voltage ratio being

$$\frac{V_{\text{out}}}{V_{\text{in}}} = \frac{1}{1001} \tag{5.4}$$

This form of divider is commonly used for high voltage applications when the small value capacitor is of the compressed gas type[2] (see figure 4.35) and where the low voltage high value unit is a good quality mica capacitor.

To obtain larger ratios a number of low value capacitors may be connected in series, hence increasing the impedance of the high voltage section. The frequency range and magnitude of voltage that can be scaled with such a divider will be limited by the leakage resistance of the capacitors and the errors introduced by stray capacitance.

Figure 5.5 shows a combined resistance — capacitance divider. This has an advantage over the simple capacitance divider, in that it may be used from d.c. up to a frequency (low MHz) which is limited by the time constant effects of the resistance—capacitance combination (see page 129).

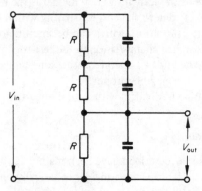

Figure 5.5. Combined resistance and capacitance divider

5.2.3 Inductive Divider

The advantages of an autotransformer voltage divider have been described in chapter 3. It will be remembered that by using a construction employing high permeability toroidal cores and interwoven windings an almost ideal transformer can be constructed whose ratio of input to output voltage is identical to the turns ratio. Using a series of autotransformers of this type, and interconnecting them using the Kelvin—Varley principle of voltage division, an a.c. voltage divider[3] may be produced which has an accuracy of 1 part in 10^7 for considerably less cost than a resistance divider of the same accuracy of ratio. However, the maximum voltage that may be applied to an inductive divider is limited by the core cross-sectional area and is typically quoted as 200 V or 0.2 x frequency, whichever is the smaller. Nevertheless the inductive divider is finding many applications in modern instrumentation, including use in some high quality digital voltmeters, where to facilitate this application d.c. to a.c. conversion must be incorporated.

5.2.4 Voltage Transformers

One of the commonest forms of voltage scaling used at power frequencies is the voltage transformer. This device is similar in construction (figure 5.6) to a power transformer (that is concentric windings on a rectangular core), operating on a very light load. Its physical size is largely dependent on the insulation required to isolate the high voltage winding from the earthed core and the steel tank.

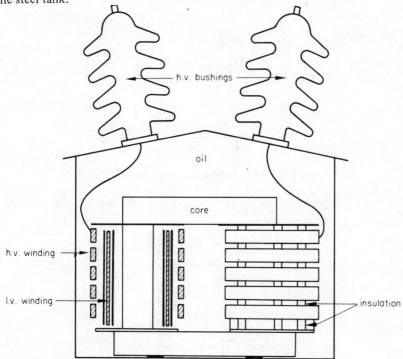

Figure 5.6. Outline drawing of a single phase double wound voltage transformer

The voltage drop due to the secondary load (or burden) and the no load (or iron) losses both contribute to the errors in ratio and phase displacement which exist between the primary (supply) and secondary (metering) terminals. One of the principle advantages of the voltage transformer (v.t.) is the electrical isolation of primary and secondary circuits allowing the high voltages of power circuits to be safely metered.

For very high voltages, 200 kV and above, the simple double wound v.t. is not a satisfactory arrangement due to the necessary physical size and cost of construction. The alternative arrangements are:

(a) *The capacitor* v.t. (figure 5.7), which consists of a capacitance divider and a double wound step down v.t. that reduces the current taken by a meter circuit. The primary of the v.t. is in series with a reactor and connected across the low voltage capacitor. The values of inductance and capacitance of the various components are selected so that resonance occurs at the frequency of operation.

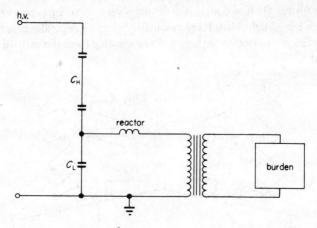

Figure 5.7. Capacitor voltage transformer

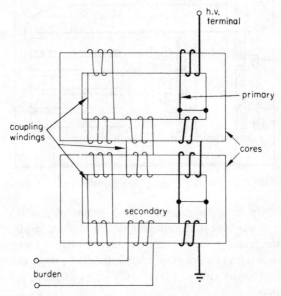

Figure 5.8. Principle of the cascade connected voltage transformer

(b)   *The cascade* v.t.[4] (figure 5.8), which breaks away from the two winding
principle in having the primary winding in several series connected
sections stepping down the high voltage in stages thereby reducing the
amount of insulation between the primary winding and the cores that are
electrically connected to the primary winding at intervals down its
length. The secondary winding is only wound on to the lowest potential
core, but coupling windings must be incorporated between each limb
of the core to ensure a balance of ampere turns in each stage when

current is flowing in the meter or burden circuit. These coupling windings also assist in reducing leakage reactance between the primary and
secondary circuit to a minimum. The complete cascade v.t. is built into
a porcelain housing; a unit for operation with a primary voltage of
400 kV would have 6 stages.

It has been stated above that the secondary output voltage will be slightly
different both in magnitude and phase from the desired or ideal value. This is of
extreme importance in metering, for if the power supplied to a load is measured
by a circuit incorporating a v.t., errors in it will affect the power meter readings
and result in a consumer being incorrectly charged. Thus it is important to have
a knowledge of these errors so that compensation or allowance may be made for
them.

Consider the simplified phasor diagram for a v.t. as shown in figure 5.9. The
difference in magnitude between the primary voltage (V_p) and the rated transformation ratio (k_n) multiplied by the secondary voltage (V_s) depends on the

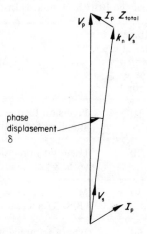

Figure 5.9. Simplified phasor diagram of a voltage transformer (not to scale)

primary current and the total effective winding impedance (Z_total) of the
transformer. Thus it is desirable that these quantities are kept to a minimum.

The British Standard on voltage transformers (B.S. 3941 : 1965)* defines the
voltage (ratio) error as

$$\frac{(k_n V_\text{s} - V_\text{p}) \, 100}{V_\text{p}} \text{ per cent} \qquad (5.5)$$

where V_p is the actual primary voltage and V_s the actual secondary voltage when
V_p is applied under the conditions of measurement.

*see note with table 5.1

Further inspection of figure 5.9 will show a small phase displacement (δ) between V_p and V_s and in B.S. 3941 this is defined as the phase error which, by convention, is considered to be positive if the phasor of V_s leads the phasor of V_p. It is apparent also from the phasor diagram that the magnitude and the phase angle of the primary current I_p will affect both the ratio and phase angle errors[2]. Now I_p has two components, I_0 the no-load current of the transformer, and I_s' the referred secondary or load current. Therefore, the load or burden connected to the secondary terminals will affect the v.t. errors. Thus a v.t. rating plate will stipulate primary and secondary voltages and the rated burden. This latter quality will be in VA (that is product of secondary voltage and current) and indicate the output at rated voltage. The phase angle of the secondary load on a v.t. will normally be very small since a voltmeter or the voltage coil of a watt-meter is predominantly resistive. However, should an inductive load be applied to

Table 5.1. *Extracted from B.S. 3941:1965: Specification for voltage transformers** (Reproduced by permission of the British Standards Institution)

Class	0.9 to 1.1 of rated voltage 0.25 to 1 of rated burden at u.p.f.		Application
	Voltage error ±	Phase error ±	
AL	0.25%	10'	Precision testing and measurements or as standard for testing other v.ts
A	0.5%	20'	Used with precision indicating instruments (B.S. 89 accuracy)
B	1.0%	30'	Meters of precision or commercial grade (B.S. 37 accuracy)
C	2.0%	60'	Where accuracy is less important than above, for example synchronising
D	5.0%	–	Ratio relatively unimportant, for example reversing transformer for synchronising
	0.05 to 0.9 of rated voltage		
E	3.0%	120'	Where only small limits of error are permissible over extended voltage range
F	5.0%	250'	Where larger limits are permissible than in E

*At the time this book was printed, B.S. 3941:1965 was under revision, the new edition (to align with B.S. 3938, see page 166) being due for publication early in 1974.

a v.t. the ratio error will be increased and the phase error, which is normally negative, may become positive.

Golding and Widdis[2] give graphs relating the effect of burden and applied voltage variations on ratio and phase errors, for a voltage transformer.

The determination of voltage (ratio) and phase errors may be performed either as an absolute measurement or in terms of the known errors of a high quality v.t.[2,5]. The classification of v.ts by their magnitude of error is stipulated in B.S. 3941 as shown in table 5.1.

5.3 CURRENT SCALING

The majority of current measuring instruments are only capable of measuring directly milliamperes or at the most a few amperes, the exceptions being some dynamometer ammeters and moving iron ammeters which are specially constructed to measure currents of the order of 100 A. Since currents of several thousands of amperes are used in many applications current scaling and voltage scaling are of equal importance.

5.3.1 Current Shunts

Current shunts have already been mentioned in connection with moving coil ammeters (see page 9), and the use of a potentiometer, both of which with the aid of shunts may be used for measuring large d.c. currents. The use of current shunts is however not limited to d.c. providing that a shunt used for a.c. current measurements is nonreactive. Its principal disadvantages are power consumption and the fact that the metering current must be operated at the same potential to earth as the current carrying line and therefore limits its use to low voltage applications[6,7,8].

A current shunt is in general a four terminal resistance, figure 5.10, the current to be measured being connected to the current terminals. Situated at points remote from the current terminals, so that they are unaffected by any heating which may occur at the current terminals, the potential terminals are arranged so that an accurately known resistance and voltage drop exists between them.

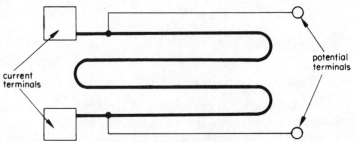

Figure 5.10. A four terminal current shunt

Applications

The application of current shunts lies in the extension of ammeter range (where the volt drop across the shunt drives a proportion of the current through the ammeter); measurement of current by potentiometer; recording of current wave-forms and the measurement of low voltage power.

Universal shunt

This is a device that may be used with a galvanometer to reduce its sensitivity and increase its input resistance when it is used as a null detector. A decade universal shunt is often built into a sensitive d.c. galvanometer. Figure 5.11 shows

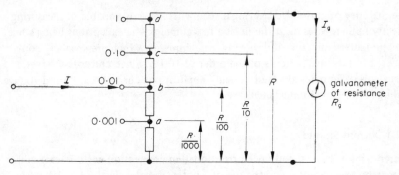

Figure 5.11. Universal shunt

the basic arrangement for a universal shunt of resistance R shunting a galvano-meter of resistance R_g. With the switch on contact a the galvanometer current

$$I_g = \frac{\dfrac{R}{1000}}{R + R_g} \cdot I \qquad (5.6)$$

and if $R \gg R_g$, $I_g = \dfrac{I}{1000}$

Similarly with the switch at point b

$$I_g = \frac{I}{100}$$

at point c, $I_g = \dfrac{I}{10}$

and at d, $I_g = I$

For a satisfactory operation of the galvanometer, R should be at least ten times greater than R_g. If R is nearly equal or less in magnitude than R_g the galvano-meter movement will become overdamped, but irrespective of this the ratio of galvanometer sensitivities for the switch positions will remain constant.

Let $R_g = R$ then:

for position a $I_g = \dfrac{I}{2000}$

for position b $I_g = \dfrac{I}{200}$

for position c $I_g = \dfrac{I}{20}$

for position d $I_g = \dfrac{I}{2}$

Thus the universal shunt increases the versatility of a galvanometer and may also enable it to be used as a multirange milliammeter. Its use is primarily for direct currents but providing its resistance is nonreactive it may be used for low frequency a.c.

5.3.2 Current Transformers

The current transformer (c.t.) overcomes the power loss and circuit isolation problems of the current shunt, but like the v.t. introduces ratio and phase displacement errors.

     The construction of a current transformer, figure 5.12, is different from that of a power transformer although the basic theory of all transformers is the same.

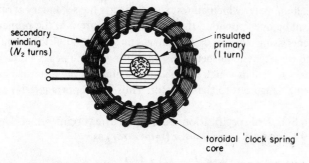

secondary winding (N_2 turns)

insulated primary (1 turn)

toroidal 'clock spring' core

Figure 5.12. Current transformer with a 'bar' primary

That is (a) the voltage induced is a transformer winding is proportional to $N\phi_m f$, where N is the number of turns in the winding, ϕ_m is the flux in the core and f the frequency of operation, and (b) the ampere turns of the windings balance. It is the latter feature which is exploited in the design of a current transformer, where to ensure that the current ratio approaches the inverse of the turns ratio the no load or magnetisation current is made as small as possible. This is attained by using a toroidal core wound from a continuous strip of high permeability steel. The primary winding is commonly a single turn (referred to as a bar primary) but may consist of a tapped winding to produce a multiratio c.t. The secondary winding, rated at either 1 or 5 A, will have many turns tightly wound on to the core.

To investigate the current error (ratio error) and phase displacement for a current transformer[2] consider the simplified phasor diagram in figure 5.13. It may be

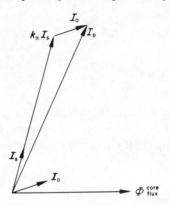

Figure 5.13. Simplified phasor diagram of a current transformer (not to scale)

seen that the difference in magnitude between the primary current I_p and the secondary current multiplied by rated transformation ratio $(I_s k_n)$, is dependent on the amount of the primary current used to energise the core and this must therefore be kept to a minimum. It should be noted that a c.t. is operated with a load consisting of an ammeter which almost short circuits the secondary terminals; should this load be open circuited the whole of the current in the primary winding will become energising current, causing magnetic saturation of the core (which is detrimental to its magnetic properties), and causing a large peaky voltage to occur at the secondary terminals which may result in failure of the interturn insulation, apart from being dangerous to the operator. Thus it is important *never* to open circuit a c.t.

The British Standard specification relating to current transformers (B.S. 3938:1973) defines *current error (ratio error)* as

$$\frac{(k_n I_s - I_p) \times 100}{I_p} \text{ per cent} \qquad (5.7)$$

Where k_n is the rated transformation ratio, I_s is the secondary current and I_p the primary current. *The phase displacement* is defined as; The displacement in phase between the primary and secondary current vectors (phasors), the direction of the phasors being so chosen that the angle is zero for a perfect transformer. The phase displacement is said to be positive when the secondary current phasor leads the primary current phasor, and negative when it lags behind the primary current phasor; it is usually expressed in minutes or in centiradians.

A current transformer connected in the supply line to a power load will be required to operate over current values from no load ($I_1 = 0$) to full load (I_1 = rated value); it is therefore important to have some knowledge of how the variation of primary current will affect the current and phase displacement errors. Figure 5.14 indicates the characteristic shape of these variations, the increasing slope at low currents being due to the changes in permeability of the core material for very low flux densities.

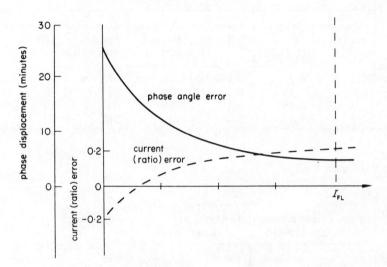

Figure 5.14. Error characteristics of a current transformer

The *burden* of a c.t. is the value of the impedance of the secondary circuit expressed in ohms (or in volt-amperes at the rated current) at the relevent power factor; a common value is 15 VA, which for a c.t. with a secondary rated at 5 A requires that the secondary terminal voltage is 3 V and the external impedance of 0.6 Ω, for operation at correct burden.

At very high voltages, the insulation of the current carrying conductor from the measuring circuits becomes an expensive problem. At 750 kV a successful solution has been the use of a cascade connected c.t. alternatively a system undergoing trials in the U.S.A. uses a coaxial shunt to modulate an r.f. signal that is transmitted from the shunt (in the h.v. line) to receiving equipment on the ground, thus overcoming the insulation problems. This type of system has a severe limitation in its power output which must be amplified to operate relays, etc.

As with v.ts the classification of c.ts (B.S. 3938) is decided by the magnitude of error and the extract from B.S. 3938 in table 5.2 shows the usage of c.ts with various tolerances.

Table 5.2. Extracts from B.S. 3938:1973: Specification for current transformers
(Reproduced by permission of the British Standards Institution)

Class	± percentage current (ratio) error at percentage of rated current shown below			± phase displacement at percentage of rated current shown below minutes		
	10 up to but not incl 20	20 up to but not incl 100	100 up to 120	10 up to but not incl 20	20 up to but not incl 100	100 up to 120
0.1	0.25	0.2	0.1	10	8	5
0.2	0.5	0.35	0.2	20	15	10
0.5	1.0	0.75	0.5	60	45	30
1	2.0	1.5	1.0	120	90	60

B. Limits of error for accuracy
Class 3 and Class 5

Class	± percentage current (ratio) error at percentage of rated current shown below	
	50	120
3	3	3
5	5	5

Note: Limits of phase displacement are not specified for Class 3 and Class 5.

C. Selection of class of accuracy of measuring current transformers

Application	Class of accuracy
(1) Precision testing, or as a standard for testing other current transformers	0.1
(2) Meters of precision grade in accordance with B.S. 37	0.2
(3) Meters of commercial grade in accordance with B.S. 37	0.5 or 1.0
(4) Precision measurement (indicating instruments and recorders)	0.1 or 0.2
(5) General industrial measurements (indicating instruments and recorders)	1 or 3
(6) Approximate measurements	5

5.4 ATTENUATORS

The term attenuator is often loosely used to denote a device which reduces the voltage and/or power conducted between the circuits connected to its input and output terminals; such an interpretation of the term may be applied to all the devices discussed so far in this chapter. A purer interpretation restricts the use of the term to apply only to those dividers which are constructed so that, as well as providing the reduction in voltage or power, the impedance of the divider is matched to the input and output circuits, and for a multiratio attenuator these impedances are constant irrespective of ratio setting.

5.4.1 Resistance Attenuator

Figure 5.15 shows a simple form of resistance attenuation pad for which $R_s = R_{in}; R_L = R_{out}$ and V_{in} is the input voltage and V_{out} the output voltage.

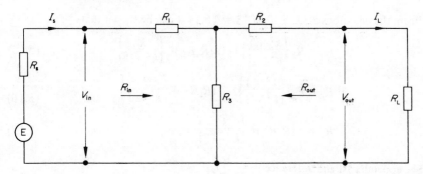

Figure 5.15. Resistance attenuator pad

It should be noted that the attenuation (A) is normally quoted as a power ratio in terms of decibels, that is

$$A = 10 \log_{10} \frac{P_s}{P_L} \text{ dB} \tag{5.8}$$

$$\text{or} \quad A = 20 \log_{10} \frac{V_{in}}{V_{out}} \text{ dB (providing } R_s = R_L) \tag{5.9}$$

Alternatively, the attenuation may be expressed in nepers (Np) when

$$A' = \log_e \frac{I_s}{I_L} \text{ Np} \tag{5.10}$$

This latter expression is commonly used in theoretical work. Now if $R_s = R_L$, then the attenuation in dB is

$$A = 20 \log_{10} \frac{I_s}{I_L} \text{ dB} = [8.686 \times \text{(attenuation in Np)}]$$

Conversely attenuation in Np = 0.1151 x (attenuation in dB).

For the attenuator pad shown in figure 5.15 let the attenuation be $20 \log_{10} k$, where

$$k = \left(\frac{P_s}{P_L}\right)^{\frac{1}{2}} = \left(\frac{I_s^2 \cdot R_s}{I_L^2 R_L}\right)^{\frac{1}{2}} \tag{5.11}$$

now $\quad R_{in} = R_s = R_1 + \dfrac{R_3 (R_2 + R_L)}{R_2 + R_3 + R_L}$. (5.12)

and $\quad R_{out} = R_L = R_2 + \dfrac{R_3 (R_1 + R_s)}{R_1 + R_3 + R_s}$ (5.13)

also $\quad \dfrac{I_s}{I_L} = \dfrac{R_2 + R_L + R_3}{R_3}$ (5.14)

From the equations 5.11, 5.12, 5.13 and 5.14 it can be shown that

$$R_1 = R_s \left\{\frac{k^2 + 1}{k^2 - 1}\right\} - 2 (R_s R_L)^{\frac{1}{2}} \left\{\frac{k}{k^2 - 1}\right\} \tag{5.15}$$

$$R_2 = R_L \left\{\frac{k^2 + 1}{k^2 - 1}\right\} - 2 (R_s R_L)^{\frac{1}{2}} \left\{\frac{k \; k}{k^2 - 1}\right\} \tag{5.16}$$

$$R_3 = 2 (R_s R_L)^{\frac{1}{2}} \left\{\frac{k}{k^2 - 1}\right\} \tag{5.17}$$

See appendix III and reference 9.

Symmetrical T

If the load and source impedance are equal the T attenuator will become symmetrical and

$$R_1 = R_2 = R_L \left(\frac{k - 1}{k + 1}\right) \tag{5.18}$$

and $\quad R_3 = \dfrac{2Rk}{k^2 - 1}$ (5.19)

If the attenuator pad is required only for matching purposes it will be arranged to have minimum attenuation and under these conditions $R_2 = 0$ resulting in an L type attenuator (figure 5.16) for which

$$R_1 = [R_s(R_s - R_L)]^{\frac{1}{2}} \tag{5.20}$$

and $\quad R_3 = \left(\dfrac{R_s R_L^2}{R_s - R_L}\right)$ (5.21)

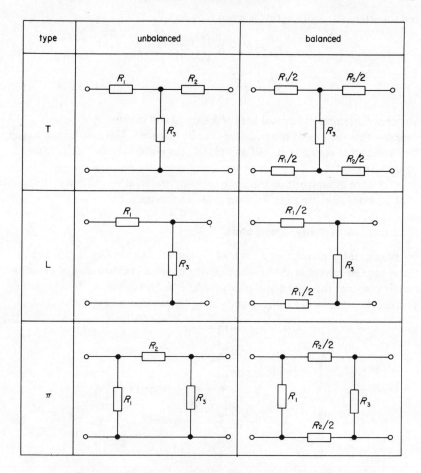

Figure 5.16. Balanced and unbalanced attenuators

Another common form of attenuator is the π type (figure 5.16) which, when used for matching, requires

$$R_1 = R_s \left\{ \frac{k^2 - 1}{k^2 - 2k \left(\dfrac{R_s}{R_L}\right)^{\frac{1}{2}} + 1} \right\} \tag{5.22}$$

$$R_2 = \frac{(R_s R_L)^{\frac{1}{2}}}{2} \left\{ \frac{k^2 - 1}{k} \right\} \tag{5.23}$$

$$R_3 = R_L \left\{ \frac{k^2 - 1}{k^2 - 2k \left(\dfrac{R_s}{R_L}\right)^{\frac{1}{2}} + 1} \right\} \tag{5.24}$$

and for attenuation alone, that is when $R_s = R_L$

$$R_1 = R_3 = R \left\{ \frac{k+1}{k-1} \right\} \tag{5.25}$$

$$\text{and} \quad R_2 = R \left[\frac{k^2 - 1}{2k} \right]. \tag{5.26}$$

The above attenuators have all been unbalanced and in some applications it is desirable that the legs of the circuit should be balanced. This is simply arranged by dividing the series arm in half and placing the equal halves in each leg (see figure 5.16).

Resistance attenuators of this form may be used from d.c. to several hundreds of MHz. For higher frequencies wave guide devices are used[10].

5.4.2 Instrument Range 'Attenuators'

Instrument range dividers are commonly called attenuators but in many cases this is not the correct term, for although the input impedance may be approximately constant the impedance presented internally will vary with range setting. Figure 5.17 shows a typical resistance range 'attenuator' and in figure 5.18 an

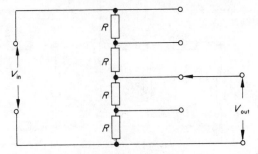

Figure 5.17. Simple resistance range 'attenuator'

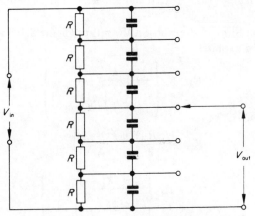

Figure 5.18. Compensated range 'attenuator'

$R-C$ arrangement is shown. This latter form of range 'attenuator' has a useful feature in that the resistance and capacitance values of the circuit may be selected to maintain constant phase shift and frequency equalisation (see page 134), it thus finds application in some oscilloscope range selection circuits. Range 'attenuators' are commonly calibrated in steps of 1, 2, 5, 10, etc., giving approximately logarithmic steps in sensitivity, but the calibration is usually in voltage and not in decibel steps.

5.5 OPERATIONAL AMPLIFIERS

The use of amplifiers in instrumentation is extensive, and in such an application they can be thought of as 'black boxes' or building bricks, their function being to operate in some manner on the signal from the measurand. This operation may simply be to increase the magnitude of the signal or it may be to perform a mathematical function on it, for example, to differentiate or integrate the signal, other possibilities being the summation of a number of signals, or the removal (filtering) of unwanted components.

Figure 5.19 shows the accepted symbols for operational amplifiers that are known as (a) single ended and (b) differential input. Both types of amplifier may

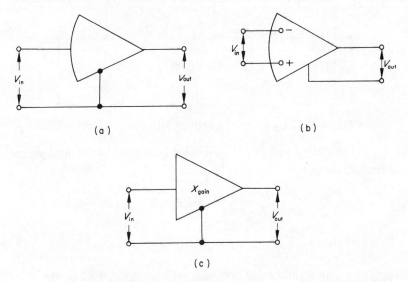

Figure 5.19. Operational amplifier symbols. (a) Single ended; (b) differential; (c) single ended with incorporated feedback circuit

either be constructed as an integrated circuit or assembled from discrete components and 'potted' in resin. In either case, as an operational amplifier, it will have a high gain, typically 10^6, it will invert (there will be a reversal of polarity between input and output), and it will have a large input impedance, typically

100 kΩ. It should also have a bandwidth which extends from d.c. upwards, together with low offset and drift properties. The symbol in figure 5.19c represents an operational amplifier with feedback.

The gain of an amplifier is the ratio V_{out}/V_{in}; thus, for an operational amplifier this should be a large negative number. However, conventionally the gain is considered as the modulus of the voltage ratio and is therefore positive. that is

$$\text{Gain} = \left| \frac{V_{out}}{V_{in}} \right| = G \tag{5.27}$$

To enhance the stability and the precision of the gain of an operational amplifier, whose nominal gain may have an appreciable specification tolerance, for example 20 per cent, lack linearity, and be temperature dependent, a proportion of the output is fed back to the input. To use an operational amplifier without feedback in an instrumentation chain would require continual recalibration of the system, but by using negative feedback on the amplifier the gain may be made largely dependent on stable passive components

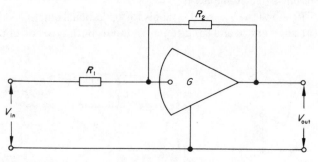

Figure 5.20. Operational amplifier with resistive feedback

and independent of the gain of the operational amplifier. For the arrangement in figure 5.20 the gain is

$$\frac{V_{out}}{V_{in}} = \frac{GR_2}{GR_1 - R_1 + R_2} \quad \text{and since } GR_1 \gg (R_2 - R_1)$$

$$\text{Gain} = \frac{R_2}{R_1} \tag{5.28}$$

The amplifier with feedback will have an input impedance of R_1 Ω, and an output impedance of a few milliohms, that is r_{out}/G at unity gain, r_{out} being the output impedance of the amplifier without feedback (typically 10 Ω to 1 kΩ). Another important effect of the use of feedback, in conjunction with the high gain of the operational amplifier, is that the input terminal is nominally maintained at earth potential, that is a few microvolts of input may result in the output reaching its maximum level, and so it is seen that the self adjusting nature of negative feedback always holds the potential of the input terminal within

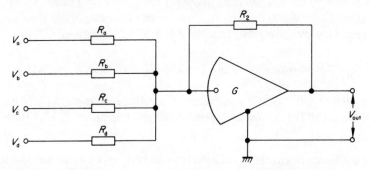

Figure 5.21. Operational amplifier used as a summing junction

a few microvolts of earth potential, no matter how the amplifier supply voltage or the values of R_1 and R_2 are varied, providing that (a) the output of the amplifier does not saturate, or (b) the upper frequency limit of the amplifier is not exceeded.

Summing amplifier

The proximity of the input terminal to zero potential is termed a virtual earth and this facet is utilised when an operational amplifier is functioning as a summing amplifier. For the arrangement shown in figure 5.21

$$V_{\text{out}} = -R_2 \left\{ \frac{V_a}{R_a} + \frac{V_b}{R_b} + \frac{V_c}{R_c} + \frac{V_d}{R_d} \right\}$$

or if $\quad\quad R_2 = R_a = R_b = R_c = R_d$

$$V_{\text{out}} = -(V_a + V_b + V_c + V_d) \quad\quad\quad\quad (5.29)$$

that is the sum of the input voltages but with a reversal of polarity.

Attenuated feedback

A further variation of the use of resistive feedback is illustrated in figure 5.22, where the 'attenuation' of feedback is a means of using low to medium value resistors while retaining a high stable gain; for example a situation in which a gain of 1000 is desired with an input impedance greater than 5 kΩ. If the circuit

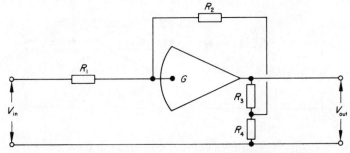

Figure 5.22. Operational amplifier with 'attenuated' feedback

in figure 5.20 were used, it would be necessary for R_2 to have a value of 5 MΩ, a value of resistance that is difficult to obtain with a tolerance of less than ± 5 per cent. However, using the circuit of figure 5.22 for which gain

$$\approx \frac{R_2}{R_1} \left\{ \frac{R_3 + R_4}{R_4} \right\}$$ providing $R_3 \ll R_2$ and with $R_2 = 100$ kΩ; $R_3 = 1$ kΩ;

and $R_4 = 20.4$ Ω. ($R_1 = 5$ kΩ) these values of resistance are more likely to be stable than the 5 MΩ of the first arrangement. See also references 11, 12, 13 and 20.

Computing amplifier

So far the feedback circuits considered have all been resistive; the introduction of reactance (normally capacitive) into the circuit enables other operations to be performed. Two such applications of major importance in the use of operational amplifiers are:

(a) as an integrator, figure 5.23a, where

$$V_{out} = \frac{1}{R_1 C} \int V_{in} \cdot dt \tag{5.30}$$

The switch is included in the circuit to ensure that there is zero stored charge in

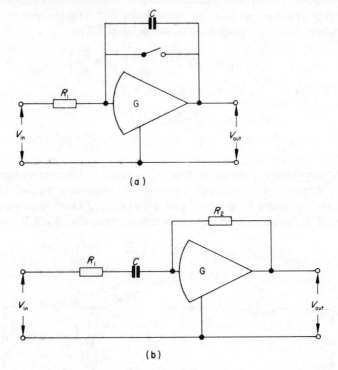

(a)

(b)

Figure 5.23. Feedback circuits to perform mathematical functions. (a) Integrator; (b) differentiator

the capacitor C at time $t = 0$. If R_1 is in Ω, C in F, and their product is 1 ΩF (one second) then the circuit is termed a 'unity integrator'.

(b) as a differentiator, figure 5.23b, where

$$V_{out} = -R_2 C_1 \cdot \frac{dV_{in}}{d_t} \tag{5.31}$$

The resistor R_1 has to be included to decrease the amplification of unwanted noise. For, the impedance of the capacitor will fall with increase of frequency and the system provides a high gain for high frequencies. The frequency limit for differentiation is

$$f = \frac{1}{2\pi R_1 C_1} \text{ Hz} \tag{5.32}$$

Charge amplifier

Some other useful applications of the operational amplifier in instrumentation are those associated with transducers. It will be shown that piezoelectric transducers have an ouput that is in the form of an electric charge proportional to the force on the crystal (see page 264), that is the transducer may in electrical terms be represented by a charged capacitor. If such a transducer is loaded solely with a circuit that resolves to a capacitance across the transducer output, the voltage developed at the transducer terminals is $V = Q/C_1$, where C_1 is the combined transducer and load capacitance. Connecting this combination across the input of an operational amplifier, with feedback capacitor C_2, as in figure 5.24, will result in an output voltage

$$V_{out} = \frac{C_1}{C_2} \times V_{in} = \frac{C_1}{C_2} \times \frac{Q}{C_1} = \frac{Q}{C_2} \tag{5.33}$$

The capacitor C_2 will normally be made equal in magnitude to C_1, the usefulness of this application being in the reduction of output impedance. Interposing the charge amplifier between the transducer and recording instrument enables instruments having low or medium input impedances to be used for recording the transducer's output.

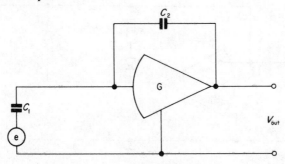

Figure 5.24. Operational amplifier used as charge amplifier

Strain gauge amplifier [11, 20]

The output voltage from a resistive strain gauge bridge is small and special
arrangements may be used. It has been found, however, that by using integrated
circuit operational amplifiers situated in the proximity of a two active arm strain
gauge bridge, inexpensive load cells of various sensitivities may be constructed
with sufficient output to drive u.v. recorder galvanometers (figure 5.25). The
most difficult problems encountered are those associated with screening and
earthing to ensure the removal of unwanted 50 Hz pickup.

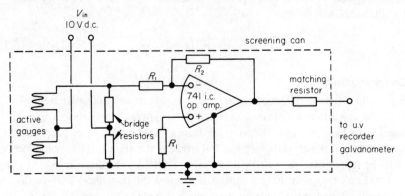

Figure 5.25. Amplification of resistance strain gauge bridge output

Ideal rectifier

In a number of applications it is necessary to determine the mean or rectified
value of a waveform (see page 21). The forward voltage drop of diodes prevents
the straightforward use of these devices if the magnitude of the signal is small.
However, if an operational amplifier is incorporated in the circuit as shown in
figure 5.26, where the diodes are in the feedback loop, this problem is effectively
overcome.

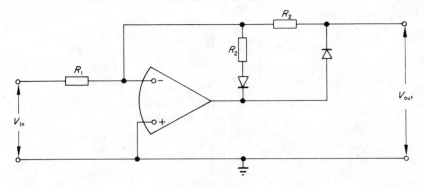

Figure 5.26. 'Ideal' rectifier circuit

Logarithmic converter

A further application of the single ended operational amplifier is as a logarithmic converter, figure 5.27. The feedback in this application is via a silicon diode, for these devices have a logarithmic relationship between voltage and current over several decades of current values. The output equation of this arrangement is

$$\log \frac{V_{in}}{R_1} = k \, [V_{out} - V']$$

where k is a constant and V' an offset voltage which must be allowed for in subsequent circuitry.

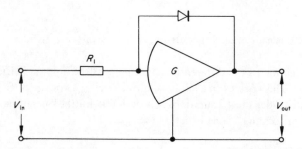

Figure 5.27. Logarithmic converter

Differential amplifier

The preceding applications have assumed that the operational amplifier was of the single ended type. Very many of the operational amplifiers that are made are manufactured with differential inputs, and to use such a device as a single ended amplifier simply requires the earthing of one of the input terminals via a resistance that has a magnitude equal to R_1 (if $R_2 \gg R_1$). The difference or differential amplifier may be used to amplify the difference between two signals. For example, figure 5.28 shows two voltage waveforms of equal frequency but unequal magnitude applied to the input of a differential amplifier. The

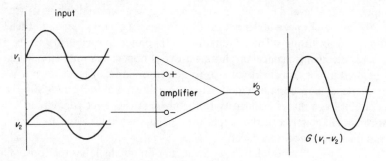

Figure 5.28. Difference of sine waves

result is an output that is an amplification of the difference between V_1 and V_2. A further illustration of the difference effect is shown in figure 5.29, where a ramp waveform is applied to one input and a sinusoidal voltage to the other.

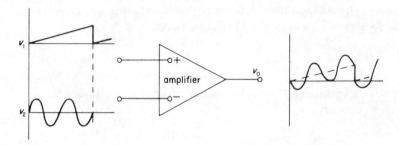

Figure 5.29. Result of applying dissimilar waveforms to a difference amplifier

One application of the differential amplifier is as the input amplifier of some oscilloscopes. This enables small voltage differences at a voltage to earth to be observed, and under these conditions it is desirable for the operational amplifier to have a high common mode rejection (see page 146).

Four quadrant multiplier

A recent innovation has been the development of integrated circuits, four quadrant multipliers[14]. These comparatively inexpensive devices facilitate a great deal of simplification in signal conditioning, for with the addition of a few components to the basic multiplier, many functions may be performed, such as multiplication, division, modulation, linearising, phase detection, power measurement, squaring, frequency doubling, and electronic gain control.

5.6 MODULATORS

To assist in the transmission and storage or display of some signals it is advantageous to modulate the signal, that is, as part of the signal conditioning process the signal from the measurand is varied, usually in a time dependent manner. For example, the low level d.c. signal input to a sensitive electronic voltmeter (see page 29) can be chopped up (modulated) into quasi a.c., amplified, and then demodulated for display. The modulation method used in such instruments are commonly electromechanical or photoresistive choppers. Another circuit that may be used to convert direct voltages to quasialternating ones is shown in figure 5.30a, an appropriate demodulator being illustrated in figure 5.30b. This latter circuit has a marked similarity to the phase sensitive rectifier circuit discussed in chapter 1, a further application being the circuit associated with the variable reluctance transducer.

The above form of modulation results in an output which has the form of a chain of pulses the heights of which are dependent on the level of the input

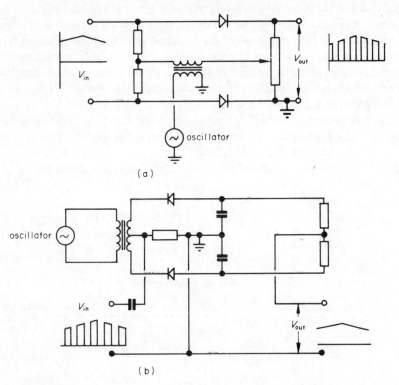

(a)

(b)

Figure 5.30. Pulse amplitude (a) modulator circuit (b) demodulator circuit

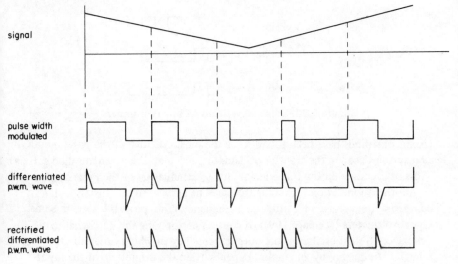

Figure 5.31. Derivation of p.w.m. and p.p.m. waveforms

signal[15], this form of modulation being termed pulse amplitude modulation (p.a.m.). An alternative to this is a technique termed pulse duration (width) modulation (p.d.m.) or (p.w.m.), in which the leading edge of the pulse occurs at fixed time intervals; the time interval between it and the end of the pulse depending on the magnitude of the signal at the time of sampling. P.P.M. (pulse position modulation) is derived from p.w.m. by differentiation and rectification; figure 5.31 shows the derivation of p.p.m. from p.w.m.

An extensively used method of representing a signal by pulses is to use a pulse code, that is a predetermined number of pulses represent discrete levels of signal, it being conventional to use a binary coding of the decimal equivalent. Figure 5.32 shows an example of pulse coding. The pulse code modulation

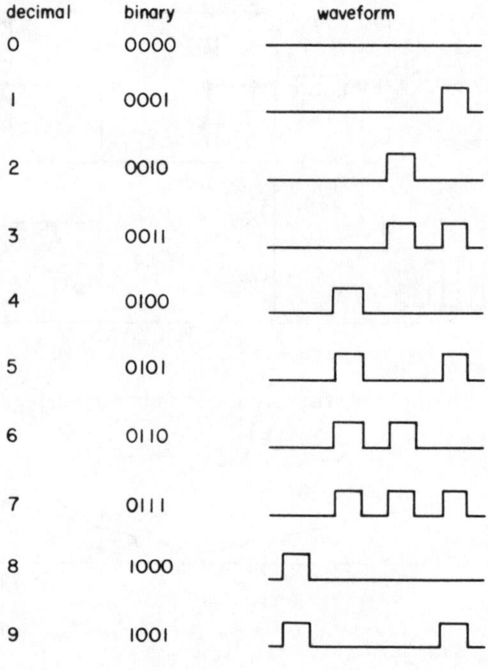

decimal	binary	waveform
0	0000	
1	0001	
2	0010	
3	0011	
4	0100	
5	0101	
6	0110	
7	0111	
8	1000	
9	1001	

Figure 5.32. Some waveforms of binary numbers

(p.c.m.) methods have been found to be the most effective of the pulse modulation systems used in the transfer of data. In particular, the p.a.m. method suffers a loss of accuracy during transmission due to attenuation of the pulse height.

Another modulation technique that may be applied to d.c. signals is that of converting the voltage magnitude to a frequency (this process is used in some digital voltmeters (see page 199)). It consists of applying the d.c. signal to a voltage controlled oscillator, the output frequency of which varies about a reference frequency by an amount dependent on the magnitude of the input signal[16].

In the above discussion the signal to be modulated has been assumed to be d.c. or to have only a slowly varying magnitude. However, for transmission purposes it is often necessary to modulate an a.c. signal; in such cases the methods used are as shown in figure 5.33, (a) being amplitude modulation (a.m.), (b) phase modulation (p.m.), and (c) frequency modulation (f.m.). The

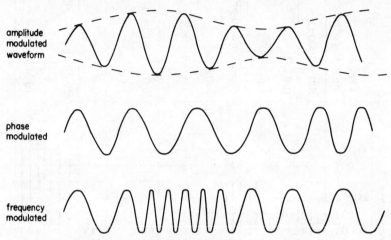

Figure 5.33. Modulation of an a.c. waveform

differences between these methods of modulation will perhaps be more easily understood if the general equation for an unmodulated waveform is considered as $V = V_m \sin(\omega t + \theta)$. Then the resultant modulated waveform may be obtained by modulating V_m, the amplitude; ω, the frequency; or θ, the phase of V with respect to a reference, it being also remembered that phase angle is equivalent to a time delay, so that phase modulation is equivalent to the variation of a time delay[17,18].

5.7 ANALOG/DIGITAL CONVERSION

The use of digital displays has undisputed advantages of clarity, reduction of operator fatigue, etc., over an analog display, and a number of analog to digital

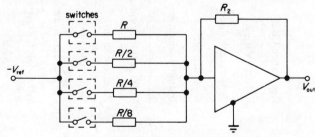

Figure 5.34. Digital to analog converter

Table 5.3. Comparison table of the various scaling methods

Type	Typical errors	Frequency Range (d.c. · 0.1 Hz · 1 · 10 · 100 · 1 kHz · 10 · 100 · 1 MHz · 10 · 100 · 1 GHz · 10)	Typical Input Impedance
Resistance Chain	0.002 → 0.1%		100 → 1000 Ω/V
Kelvin−Varley	0.001 → 0.01%		100 kΩ
Capacitance divider	0.01 → 1%		50 pF
R−C divider	0.1 → 1%		10 MΩ and 50 pF in parallel
Inductive divider	1 pt in 10^8 to 0.01%		100 kΩ
Voltage transformer	0.1 → 5%		
Resistive shunt	0.01 → 1%		0.01 Ω → 100 Ω
Universal shunt	0.1 → 1%		1 kΩ
Current transformer	0.1 → 5%		0 → 10 Ω
Attenuator/Resistance probe	0.1 → .5%		10 MΩ and 50 pF in parallel
Diode probe			10 MΩ

conversion techniques[19] are described in connection with the operation of digital voltmeters (see page 197). However, it is sometimes advantageous to convert digital signals to analog ones for display purposes, for example, presenting computer output in graphical form. The process of digital to analog $(D-A)$ conversion is illustrated in principle in figure 5.34, where the summation properties of an operational amplifier are used to form an analog voltage from digital information (see page 173). The summing resistors (R) have values in a binary progression and are connected to a fixed voltage source via switches ('flip-flop' circuits) that are controlled by the digital information. The voltage source must have a precisely known value, and the summing resistors must be stable. The scale of the output voltage will be dependent on the value of the feedback resistor R_2.

5.8 COMPARISONS

Table 5.3 shows a comparison of the frequencies over which it is normal to use the signal conditioning devices described in this chapter.

REFERENCES

1   Jan Slyper. *Advanced D.C. Calibration Techniques.* E.E.E. European Seminar (April 1969)

2   E. W. Golding and F. C. Widdis. *Electrical Measurements and Measuring Instruments* Pitman, London (1963)

3   J. J. Hill and A. P. Miller. 'A Seven Decade Adjustable Ratio Inductively Coupled Voltage Divider with 0.1 ppm Accuracy'. *Proc. I.E.E.* **109B** (1962)

4   'Cascade Voltage Transformers'. *I.E.E. Electronics and Power.* 470—471 (December 1967)

5   B. D. Jenkins. *Introduction to Instrument Transformers,* Newnes, London (1967)

6   F. E. Terman and J. M. Petit. *Electronic Measurements,* McGraw-Hill, New York (1952)

7   F. K. Harris. *Electrical Measurements,* Wiley, New York (1962)

8   J. H. Parks. 'Shunts and Inductors for Surge Current Measurements'. *N.B.S. J. Research.* **39**, 191 (1947)

9   The Royal Signals. *Handbook of Line Communications* Vol. 1. H.M.S.O., London (1947)

10   M. Sucher and J. Fox (Eds.) *Handbook of Microwave Measurements,* Wiley, New York (1965)

11   *Notes on Operational Amplifiers.* Fenlow Electronics Ltd

12   R. Morrison. *D.C. Amplifiers in Instrumentation,* Wiley, Interscience, New York

13   G. B. Clayton. *Operational Amplifiers,* Butterworth, London (1971)

14   *Evaluating, Selecting, and Using Multiplier Circuit Modules for Signal Manipulation and Function Generation.* Analogue Devices Ltd (1970)

15   R. H. Cerni and L. E. Foster. *Instrumentation for Engineering Measurements*, Wiley, New York (1962)

16   F. Bombi. 'High Performance Voltage to Frequency Converter has Improved Linearity, *Electronic Engineering* (Dec. 1970)

17   S. J. Mason and H. J. Zinnerman. *Electronic Circuits, Signals and Systems*, Wiley, New York (1960) and Chapman and Hall, London (1960)

18   S. Stein and J. Jay-Jones. *Modern Communication Principles*, McGraw-Hill, New York (1967)

19   *Logic Handbook*. Digital Equipment Corporation (1968)

20   R. J. Isaacs, Optimising Op. Amps. *Wireless World* (April 1973)

6

Digital Instruments

Digital instruments display the measured quantity in discrete numerals thereby eliminating the parallax error, and reducing the human errors associated with analog pointer instruments. In general digital instruments have superior accuracy to analog pointer instruments, and many incorporate automatic polarity and range indication which reduces operator training, measurement error, and possible instrument damage through overload. In addition to these features many digital instruments have an output facility enabling permanent records of measurements to be made automatically.

Digital instruments are, however, usually more expensive than analog instruments. They are also sampling devices, that is the displayed quantity is a discrete measurement made, either at one instant in time, or over an interval of time.

6.1 DISPLAY METHODS

The method of displaying the numerical digits of the unknown quantity are principally as listed below.

(a) *Mechanical drum or disc.* Digits engraved or printed on to a disc or drum are displayed in a window by the operation of either a manual or an electromechanical system. Examples of this are the solenoid operated counter, the modern electric energy meter and the hodometer (miles travelled) fitted in motor cars.

(b) *Neon tubes.* These have a variety of forms[1], the commonest three being:
(i) the cold cathode numerical indicator or digital display tube (figure 6.1) in which the numerals 0–9 are formed as electrodes and applying a potential to the appropriate electrode causes it to glow and hence form an indication;
(ii) the cold cathode transfer tube, in which a moving glowing spot rotates one position for each input pulse, the spot having ten positions labelled 0–9. Whilst this device has advantages in the simplicity of

associated circuitry (the tube itself performing the counting) it is not truly a digital display and is thus falling from use in digital instruments. (iii) the segmented display (figure 6.2) in which seven or sixteen neons are arranged in a figure of eight pattern, the appropriate neons being energised to display the numerals 0–9.

(c)   *Incandescent displays.* These may either project the numerals from a lamp and lens system on to a window; or take the form of a stack of transparent plates each engraved with a number that may be edge lit by its own lamp.

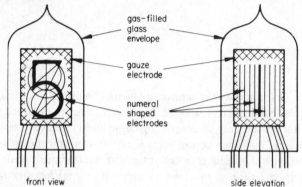

front view   side elevation

Figure 6.1. Cold cathode numerical indicator tube. (a) Front view; (b) side elevation

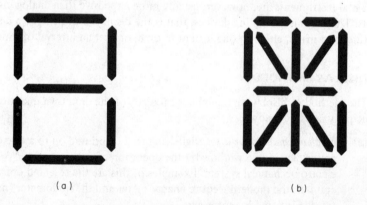

(a)   (b)

Figure 6.2. Segmented display numerals. (a) Seven segment; (b) sixteen segment

(d)   *Solid state*[3,9,11]. This is a development of integrated circuit technology and results in a pattern of 27 or 35 light emitting diodes which, like the segmented neon display, may be energised to obtain the display numerals (figure 6.3a and b).

(e)   *Liquid crystals.* These are organic materials that at room temperatures are in the mesomorphic state (liquid but with the ordered structure of a solid). A character cell is made by trapping a thin layer of liquid crystal

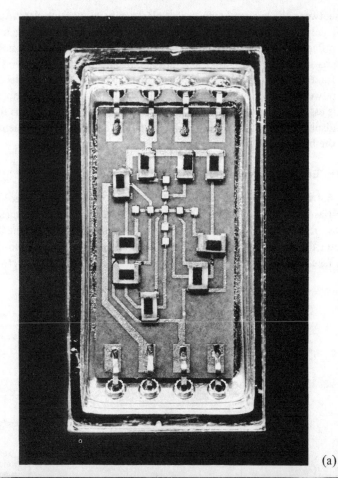

(a)

(b)

Figure 6.3. (a) A plus sign formed by light emitting diodes; (b) Solid state numerical indicators (four 3-digit modules). The numerals are formed by selecting appropriate diodes from an array of 27 (Courtesy Hewlett Packard)

between the layers of transparent electrically conducting glass. Applying an electrical field to the transparent crystal causes it to become milky and reflecting. The shape of the character is formed by etching the conducting coating on the glass into the required pattern, for example 7 bar shape for a numerical display (see references 11 and 12). While the electrical power required for operation is small (1 mW per character), ambient light is essential, the display being invisible in darkness, although this form of display has the advantage that the greater the intensity of ambient light the brighter the display appears.

6.2 DIFFERENTIAL VOLTMETERS

Figure 6.4 shows a schematic diagram of a differential voltmeter. There are similarities between it and the potentiometer (see page 77) in which measurement depends on the equalising of a known voltage with an unknown. In the differential voltmeter the reference voltage is large and compared through a Kelvin–Varley divider (see page 153) with the voltage to be measured. The null detector used is

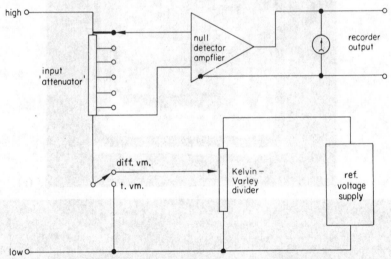

Figure 6.4. Simplified diagram of a differential voltmeter

normally a transistor voltmeter (t.vm.) with a high input impedance (typically 10^{10} Ω). To obtain an indication of the input voltage magnitude the t.vm. may by connected through an input divider and used independently of the differential voltage circuits. Having established the order of the unknown voltage, the instrument's mode of operation may be switched from t.vm. to diff. vm.[7], and with a suitable range selected, the decades of the unknown voltage can be determined in sequence. The sensitivity of the null detector being increased as the consecutive decades of voltage magnitude are established. The internal high voltage reference is maintained constant by using zener diodes or a voltage reference amplifier (see page 203) in its circuitry, and stabilities of the order of 0.003 per cent are quoted in differential voltmeter specifications. The output facility from a differential voltmeter consists in taking the voltage applied to the null indicating pointer instrument to a pair of terminals so that this voltage may be connected to a pen recorder, enabling variations between the unknown voltage and the voltage set on the dials to be recorded.

A.C. differential voltmeters are essentially the same as d.c. units except that the a.c. input signal is converted to d.c. before it reaches the null detector amplifier, the reference voltage being d.c. The conversion technique used may be either mean or true r.m.s. sensing; if of the latter form, and utilising an evacuated thermocouple device, errors in measurement less than ± 0.05 per cent of reading + 0.005 per cent of range over a frequency range of 30 Hz–50 kHz and a voltage range of 10 mV–1100 V irrespective of input waveform, may be attained.

Features

The main features of the differential voltmeter may be summarised as follows:

(a) Digital display (mechanical drum or disc).
(b) Very good accuracy at comparatively low cost.
(c) High stability of internal reference voltage (for example, 5 p.p.m. over 25 h).
(d) Recorder output, making it possible to obtain a continuous record of variations in input level.
(e) May have selfalignment and calibration facility.
(f) Some units may be operated from an internal battery, thus reducing isolation problems (see page 139).
(g) D.C. only; a.c. only; and a.c.–d.c. versions are available, all having very high input impedances.

Applications

Voltage measurement

The differential voltmeter may be used for making measurements of voltage when a high degree of accuracy is required. Its principal advantage in this application is the output facility enabling permanent records to be made of variations in the level of the input voltage. Since it is a voltmeter with inherently small errors it can also be used to calibrate other instruments (see page 221).

Current measurements

Since the differential voltmeter has a very high input impedance (tending to infinity at balance) it may only be used to measure current as a voltage drop across a known resistor, but in this way it can be used to record variations in current level.

Ratio measurements

A number of differential voltmeter models incorporate the facility for direct measurement of ratio. This is performed by disconnecting the voltmeter's Kelvin–Varley divider from the internal reference supply and reconnecting it to the external supply (d.c.) across the divider under test (see figure 6.5). Adjustment of the

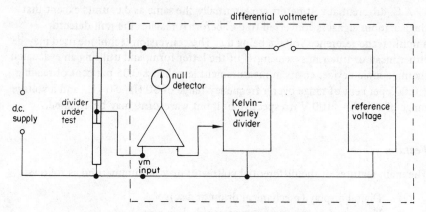

Figure 6.5. Use of a differential voltmeter for ratio measurements

voltmeter decade switches is then made to give a direct reading of ratio when zero is obtained on the null meter.

D.C. amplifier

This is a straightforward use of the output feature, that is the null detector may be utilised as a narrow band d.c. amplifier.

6.3 COUNTERS

Electronic counters are digital instruments that can measure the number of pulses in a known time interval, or alternatively the time interval between pulses. Their accuracy of measurement is largely dependent on an internal oscillator and they are marketed in variations of frequency range and versatility.

Principle of Operation

A digital counter may be considered to consist of a number of operational blocks that may be interconnected to perform various functions.

Display. The display method for the majority of digital counters utilises an 'in-line' arrangement of numerical indicating tubes (see figure 6.6), or light emitting diodes. The basic method of electronic counting uses 'flip-flop' circuits, which have two stable states[1,2]. The application of a pulse to such a circuit will cause it to change from one state to the other. The usage of such circuits is compatible with

Figure 6.6. A counter timer with an 'in line' display formed from cold cathode numerical indicator tubes (Courtesy RACAL Ltd)

binary counting and table 6.1 tabulates some binary coded and decimal equivalents. The rapid interpretation of a binary display is a skilled operation and so binary to digital convertion, commonly by a diode matrix, must be performed within the counter[8].

The output (to drive printers or tape punch) from a digital instrument is normally a modified form of binary coding.

Table 6.1. Table of binary and decimal equivalents, and supplementary codes.

		Codes in Common Use			
Decimal	Binary	8−4−2−1	2−4−2−1	5−2−1−1	Excess−3
0	0	0000	0000	0000	0011
1	1	0 001	0001	0001	0100
2	10	0010	0010	0011	0101
3	11	0011	0011	0101	0110
4	100	0100	0100	0111	0111
5	101	0101	1011	1000	1000
6	110	0110	1100	1001	1001
7	111	0111	1101	1011	1010
8	1000	1000	1110	1101	1011
9	1001	1001	1111	1111	1100

Internal Oscillator. To determine the time over which an unknown frequency is measured, time intervals must be derived. In some inexpensive counters the power line frequency may be used for this purpose but this is a poor reference as it has a nominal tolerance of ± 0.5 Hz and variations of ± 2.5 Hz are not unknown. Therefore the majority of counters incorporate an internal, fixed frequency, quartz oscillator operating in the 1−10 MHz range. The stability of this oscillator is often enhanced by mounting it in a constant temperatureoven within the instrument, when stabilities of 1 part in 10^8 or better are obtainable. To increase the versatility of the high frequency oscillator its output is connected to a number of frequency dividers, normally arranged in decades, for example a 1 MHz oscillator connected to 6 decade dividers could produce pulses at intervals of 1 μs, 10 μs, 100μs, 1 ms, 10 ms, 100 ms, and 1 s, by taking the output from suitable points in the decade divider chain.

Main Gate. To control the interval over which pulses are applied to the display unit a gate is incorporated. This is a pulse operated switch, one command pulse opening the gate (start) and allowing the passage of count pulses; the next command pulse closing the gate and preventing the flow of count pulses.

The remaining parts of the counter are (a) trigger and pulse shaping circuits which control the voltage level and shape of pulses to the gate and to the display unit; (b) selector switches linked to (c) the logic circuits which determine the mode of operation of the counter as described under applications, the appropriate units of the display, the duration and resetting to zero of the display, and the gating interval.

Features

(a)   Digital display, and in many cases an output (usually) in BCD form.
(b)   Versatile instrument for measuring frequency and time intervals, although not every counter can be used for all the operations listed under applications.
(c)   An instrument with small errors which are dependent on: (i) the internal oscillator, and (ii) an uncertainty of ±1 digit in the least significant figure of the display.

Applications

It was stated above that the digital counter consists essentially of (a) a digital display unit, (b) a timebase or oscillator unit, and (c) a main or control gate. By using additional circuitry these component parts may be made to perform a variety of counting measurements.

Totalising

This is the simplest count function consisting of routing the pulses to be counted through the main gate which may be started and stopped either manually (by push button switches) or by some externally derived pulses (see figure 6.7). The display thus recording the total number of pulses received during the interval between the start and stop signals.

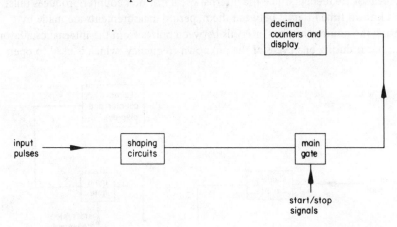

Figure 6.7. Schematic diagram of connections for totalising count

Frequency measurement

The logical step from the above section is to supply command signals to the gate at known time intervals (for example 1 s) derived from the internal oscillator, when the totalised number of incoming pulses will be a direct measurement of frequency (see figure 6.8). To measure high frequencies a shorter time interval may be used.

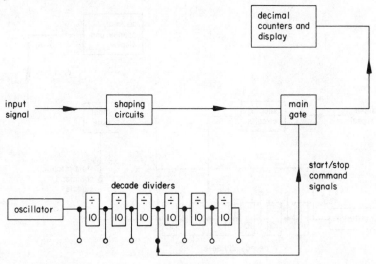

Figure 6.8. Schematic diagram of connections for frequency measurement

Period measurement

The period of a waveform may be defined as the time interval between identical points in successive cycles, for example positive going zero crossings. It is also the reciprocal of frequency. Since the internal oscillator of a counter produces pulses with a known time interval between them, period measurements are made by counting the number of time intervals between pulses from the internal oscillator which occur during one cycle of the unknown frequency, which is used to open

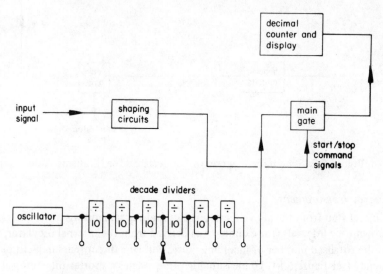

Figure 6.9. Arrangement for period measurement

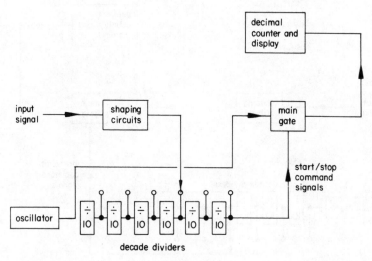

Figure 6.10. Arrangement for multiple period averaging

and close the main gate, by connecting the component parts of the counter as in figure 6.10. Period measurements enable more accurate determination of low frequency signals to be made than would be obtained using a direct frequency measurement due to the increased resolution possible, although an allowance for the trigger error must be made.

Multiple period averaging

The resolution for medium frequency signals can be increased and the trigger error decreased by routing the input signal via the decade dividing assemblies (see figure 6.10). The number of cycles of input over which the internal oscillator's pulses are counted then being increased by powers of ten.

Ratio measurements

The ratio of two frequencies is determined by using the lower frequency signal to operate the gate while the higher frequency signal is counted (see figure 6.11). This technique may be used for the conversion of transducer output signals from pulses to practical units, a number of commercially available counters incorporate selectable divider ratios to assist in this application.

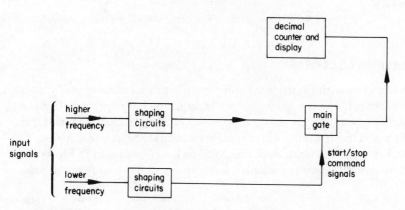

Figure 6.11. Arrangement for frequency ratio measurement

Time interval measurements

These are similar to period measurements except that the command signals to the main gate may be derived from adjustable points either on a single waveform or on two separate waveforms (see figure 6.12). The triggering polarity, amplitude, and slope may be selected independently for the open and close signals and the time interval displayed in units of μs, ms, or s.

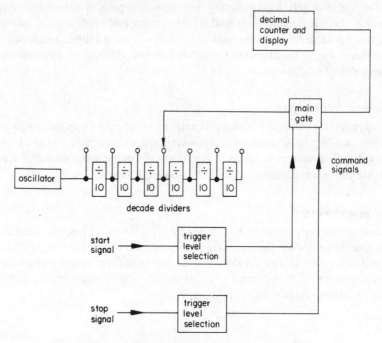

Figure 6.12. Arrangement for time interval measurements

6.4 DIGITAL CLOCKS

In digital measuring systems the sequence of events is invariably time controlled, in addition to this it may be desirable to display and/or record time in digital form. In most cases the timebase frequency for a digital clock is derived from the power line frequency but should better accuracy be desired a crystal oscillator or even atomic resonance could be used (see page 212). Figure 6.13 shows a block diagram of a digital clock deriving its timebase from the power

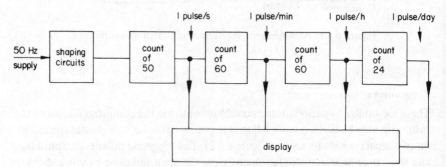

Figure 6.13. Schematic diagram of 'mains' reference digital clock

line frequency and capable of displaying, hours, minutes, and seconds on a 24 h basis[4]. As an alternative to deriving 50 pulses per second from the power line, which may have additional spurious spikes, the mains supply may be used to drive a synchronous motor coupled to a transducer producing 3000 or more pulses per second giving system command pulses and resolution of a shorter interval than 20 ms (see page 253).

6.5 DIGITAL VOLTMETERS

A digital voltmeter[8] (d. vm.) converts a sample of an unknown voltage input into a digital quantity which is then displayed in numerical form, usually by cold cathode numerical indicator tubes, but other display methods may be used. Since good resolution may be obtained by using numerical readout the accuracy with which the input voltage is measured depends largely on the errors in the analog to digital conversion technique employed in the operation of the instrument.

Principle of operation

There are five main methods used in the construction of digital voltmeters for the conversion of an analog signal to a digital one. These are:

- (a)   successive approximation or potentiometric method
- (b)   ramp or voltage to time conversion technique
- (c)   voltage to frequency method
- (d)   dual slope technique
- (e)   recirculating remainder system.

Each method has advantages of its own, making a particular conversion technique more suitable for some applications than others. Thus an appreciation of the operation of these conversion techniques is desirable.

6.5.1. Successive approximation method

This is the fastest and one of the most stable of the basic analog to digital conversion techniques. Instruments using this method work automatically in a similar manner to the operator of a normal laboratory d.c. potentiometer. In the successive approximation d. vm. (see figure 6.14) a voltage divider network, with coarse and fine steps, is connected via reed or transistor switches to a voltage comparator (the equivalent of the potentiometer operator's galvanometer), which compares the internal voltage with the unknown. The output of the comparator feeds the logic circuits which control the steps on the voltage divider network. A measurement sequence usually selects the largest steps of the internal voltage first, the magnitude of the steps decreasing until the null point is reached.

   The high speed of measurement possible with this technique only applies if the unknown voltage is noise free. If it is not, filters must be fitted and the speed of operation is very much reduced.

The errors associated with a d. vm. employing this method of conversion will depend on (a) the resolution of the comparator, (b) the precision of the voltage divider network which is normally resistive but in some good quality instruments

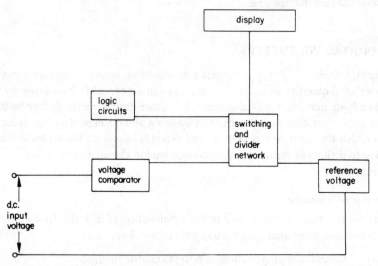

Figure 6.14. Simplified schematic diagram of a successive approximate d. vm.

may be an inductive divider, and (c) the stability of the reference voltage which may be derived from a standard cell or a zener diode circuit.

The overall performance of this type of converter may also be limited by the divider switch characteristics for although solid state switching may be accomplished at a higher speed than electromechanical switching the latter has a superior on–off characteristic.

6.5.2. Ramp method

This method of conversion utilises a digital counter technique. A carefully defined, internally generated, voltage ramp is fed to two voltage comparators (see figure 6.15). When the ramp voltage, which may have either a positive or negative slope, is equal to one level of the input voltage (say earth) the ground comparator emits a command (start) signal to a gate which opens and permits the passage of pulses from a crystal oscillator to a digital counter. When the level of the input voltage is equal to the ramp voltage the second comparator produces a command signal to close the gate and prevent further counting. With suitable scaling the count may be made of equal magnitude to the difference in the two levels of the input voltage. The factors affecting the errors of a d. vm. using this conversion technique are (a) the linearity of the ramp voltage, which is normally generated

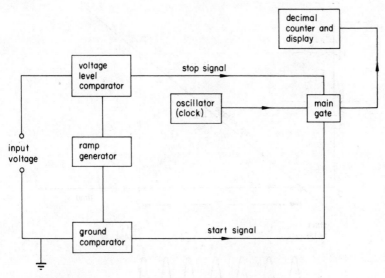

Figure 6.15. Basis of ramp convertion technique d. vm.

by using an amplifier with capacitive feedback, (b) the stability of the crystal oscillator, which commonly operates at 1 MHz and (c) the precision with which the voltage comparators can make coincidence measurements. As with instruments using the successive approximation method of conversion the measurement rate for input voltages having a noise content is reduced by the inclusion of filter circuits.

6.5.3. Voltage or frequency method

Although this method also employs a counter technique the mode of operation is fundamentally different to the ramp technique. In the voltage to frequency conversion a signal is generated such that its frequency is precisely related to the differences in the levels of the input voltage. This frequency is then counted over a fixed time interval, usually 1 cycle of power line frequency, which results in a high rejection of mains noise signal without the use of filters. A schematic diagram of the components of a voltage to frequency d. vm. is shown in figure 6.16.

The errors of a d. vm. using this technique are dependent on: (a) the accuracy and linearity of the voltage to frequency conversion, which is not as inherently stable or accurate as the successive approximation method, (b) the precision of the time interval over which the frequency measurement is made, which may be made small by using crystal control, and (c) the internal reference or calibration voltage.

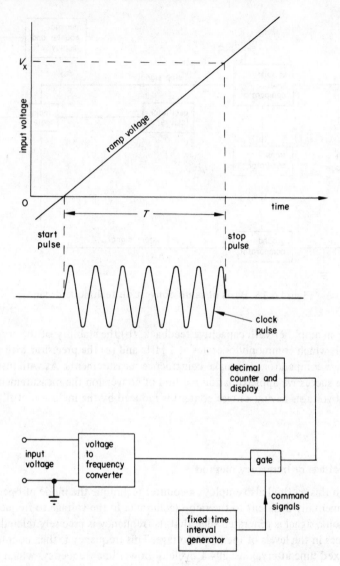

Figure 6.16. Schematic diagram for a voltage to frequency d. vm.

6.5.4. Dual slope technique[13]

In this method of analog to digital conversion an attempt is made to combine the advantages and remove the disadvantages of the two preceding methods[5], for whilst the actual measurement is a voltage to time conversion the sample time is constant and can be arranged to reject power line noise. Thus the unknown voltage is determined by a two-stage operation; the first stage of which occurs in a fixed time $T = 1$/mains frequency, during which a capacitor (operational amplifier) is charged at a rate proportional to the input voltage (see figure 6.17). At the end of time T the input to the operational amplifier is switched to a reference voltage, of opposite polarity to the input voltage, and the capacitor discharged at a constant rate giving time intervals, for pulses to flow to the clock, proportional to the magnitude of the input voltages, for example, $t_1 \propto V_1$;

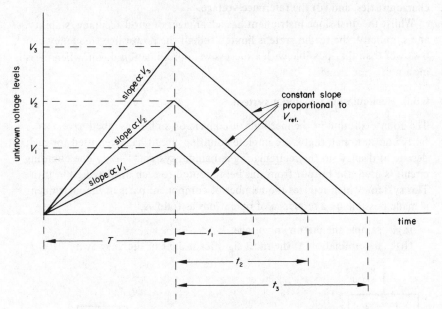

Figure 6.17. Voltage–time relationships in a dual slope d. vm.

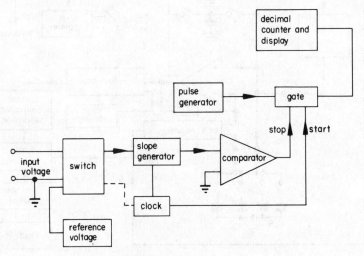

Figure 6.18. Schematic diagram of a dual slope d. vm.

$t_2 \propto V_2$, etc., figure 6.18 shows a simplified schematic diagram for a dual slope d. vm.; the clock and pulse generator signals would be derived from a common crystal oscillator operating at, say, 60 MHz for an instrument with five figure resolution. The error of reading is not however dependent on this frequency but is affected by (a) the 'input or reference' switch characteristics, (b) the voltage and leakage characteristics of the operational amplifier, (c) the comparator characteristics, and (d) the reference voltage.

Whilst the dual slope instrument has advantages of good accuracy, stability, and simplicity, the reading rate is limited to half the power line frequency which is slower than that possible with a successive approximation d. vm. when measuring a noise free signal.

6.5.5. Recirculating remainder system

The counters found in the methods of conversion so far described have been fairly conventional, that is a complete counting circuit had been used for each decade of display. In the recirculating remainder system[6, 10] only one counting circuit is used, the output from this being routed to each display digit in turn. This system, which reduces the number of component parts of the instrument, is made possible by a sequence of operations as follows;

(a)   sample the unknown voltage
(b)   determination of the most significant digit of the unknown

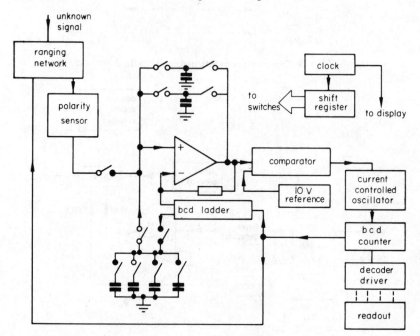

Figure 6.19. Recirculating remainder d. vm. (John Fluke)

(c)   storing the difference between the first digits worth and the unknown
      in a capacitor
(d)   amplifying this remainder by a factor of ten
(e)   determine the next most significant digit
(f)   storing the remainder.

The steps d, e and f are cycled until the magnitude of the unknown voltage is
displayed with the full resolution available.

Figure 6.19 shows a schematic diagram of this type of d. vm. which is comparatively new, makes greater utilisation of component parts, and has a strobed
display (thereby consuming less power) facilitating battery operation. Since
the same capacitors are used to store all the digits, each is determined with the
same precision. The sampling time is small (4 ms) and the conversion fast,
giving a reading rate of 40 per second. The accuracy limiting factors are: (a)
stability of the operational amplifiers which can be made of the order of a
few p.p.m., (b) the reference voltage, which is derived from a reference
amplifier – a combined zener diode and transistor i.c. having zero temperature
coefficient over normal ambient temperatures, (c) leakage and dielectric storage
of the remainder storage capacitor, which is much less of a problem in this conversion technique than in the dual slope method since here only the remainder is
stored.

Features

The principal features of digital voltmeters are as follows:

(a)   Basically a high input impedance d.c. voltmeter.
(b)   A sampling instrument.
(c)   Higher accuracies available than with analog instruments.
(d)   Higher cost than analog pointer instruments.
(e)   Numerical display giving greater resolution, increased reading speed
      and reduction of human error.
(f)   Many d. vms have an output facility enabling permanent records to be
      made on punched tape, magnetic tape or by printer.
(g)   Some d. vms have automatic polarity and range changing features which
      reduce operator training, measurement error and possible damage through
      overload.

Applications

D.C. voltage measurements

D. vms can only measure d.c. voltages, and thus all other quantities must be converted to this form before measurements can be made. The majority of d. vms
have an input range switch enabling d.c. voltages up to 1000 V to be measured.
The smallest voltage (resolution) which can be measured is commonly 10 μV
but some instruments have a resolution of 2.5 μV or even 1 μV.

Table 6.2. Comparison table of digital instruments

type	error as percent of fiducial value	maximum sensitivity	approx. input impedance	output	display	reading rate	frequency range Hz	remarks
differential voltmeter	$0.002 \to 0.1$	$0.2\ \mu V$	$10\ M\Omega \to 10^{11}\ \Omega$	difference voltage	digits engraved on disc or drum	depends on operator	d.c. and ~10 to 10k Hz	d.c.; a.c.; and a.c./d.c. models available (mean or true r.m.s. sensing). Currents measured as volt drop
digital voltmeter	$0.002 \to 0.5$	$1 \to 10\ \mu V$	$10 \to 100\ M\Omega$	bcd	numerical indicator tubes, or l.e.d.	$2/s \to 100/s$	d.c. and up to ~1M Hz	d.c. only; more expensive than diff. voltmeter but may be completely automatic. a.c. measured using a.c./d.c. converter additional cost and normally mean sensing
counter timer	± 1 digit	$1\ \mu V \to 100\ mV$	$10\ k\Omega \to 10\ M\Omega$	bcd	ditto	$1/10s \to 20/s$	depends on units (~10 to 1G Hz)	single count function and multiple count/timing function unit; available. See Digital Instruments by W. F. Waller[8]

d.c. potentiometer + range box	0.002 → 0.1	0.2 μV	→ ∞ 100 Ω/V	none	numbers engraved on disc or pointed at, rarely as clear as Diff.Vm.	depends on operator		max. voltage measurable without range box is 1.8 V. Ergonomics tend to be less convenient than diff. vm. giving slower reading rate and less convenient to use
a.c. potentiometer + range box	0.2 → 1	0.5 mV	→ ∞ 100 Ω/V	none	ditto	ditto		as for d.c.; however has the adv. that it may be used for phase meas. and the meas. of watts at low p.f.

D.C. current measurements

These must inevitably be made as a voltage drop across a known resistance.

A.C. measurements

As indicated above, to perform a.c. measurements an a.c. to d.c. converter must be inserted between the unknown a.c. voltage and the digital voltmeter. Depending on the method of conversion used, the display will be derived from a mean sensing, a peak sensing, or a true r.m.s. converter, and care must be exercised when using a d. vm. to measure a.c. quantities. It should also be noted that for a d. vm. fitted with an a.c. to d.c. converter the errors of a.c. measurement are greater than those for d.c. values.

Multifunction digital multimeters (d. mms)

The reductions in cost of digital displays and the developments in integrated circuit technology have made multifunction digital instruments economically more attractive with instruments of better performance for the same or less cost appearing at roughly yearly intervals. The functions available on the conventional multimeter are a.c. and d.c. voltage; a.c. and d.c. current; and resistance. Some more sophisticated multifunction instruments can, by making use of 'plug-in' modules, be used for such measurements as voltage ratio (a.c. and d.c.); resistance ratio (2 and 4 terminal) and scaling of transducer outputs[5,6,8] (see also chapter 11).

6.6 COMPARISONS

Table 6.2 compares the principal characteristics of the various forms of digital instruments. The d.c. and a.c. potentiometer characteristics have been added to this table since in common with the differential and digital voltmeters, they should be considered for applications in which a low error measurement of voltage is required. This table should only be considered as a general guide and a careful study of current specifications of the various instruments should be made to obtain a comprehensive picture of the characteristics of such instruments.

REFERENCES

1   R. H. Cerni and L. E. Foster. *Instrumentation for Engineering Measurements.* Wiley, New York (1962)

2   Sol. D. Prenski. *Electronic Instrumentation.* Prentice Hall, Englewood Cliffs, N.J. (1963)

3   H. C. Borden *et al.* 'Solid State Displays'. *H.P. Journal* (Feb. 1969; July 1970) and *H.P. Application Note 931.*

4   *Logic Handbook.* Digital Equipment Corporation (1968)

5   *H.P. Journal* (March 1969)

6   R. Van Saun and F. Capell. 'Serial Conversion Knocks some Stuff out of
     D. Vms.' *Electronics* (May 25th, 1970)
7   R. J. Mewett. 'The Differential Voltmeter'. *Electronic Equipment News*
     (April 1970)
8   W. F. Waller (Ed.). 'Digital Instruments'. *Electronic Data Library Vol. 1*
     Product Journal Ltd. (1968)
9   J. McDermott. 'Solid-state Optoelectronic Components put imagination in
     Engineering'. *Electronic Design.* **19**, Nos. 11 and 12 (May–June 1971)
10   R. J. Mewett. 'Conversion Techniques for D. Vms.'. *Electronic Equipment
     and News* (Nov. 1971)
11   D. W. Byatt. 'Solid-state and Liquid Crystal Displays'. *Electronics and
     Power.* 398–400 (Nov. 1972)
12   L. Harrison. 'Liquid crystal displays'. *Electron* 17–21 (Oct. 12 1972)
13   J. R. Pearce. 'Dual ramp voltmeters – basic theory' *Solartron. D. vm.
     Monograph No. 7*

7

Errors in Measurement

No measurement is perfect. However, the imperfections of a measurement must be established and they are normally quoted as the tolerance or errors of measurement in relation to a standard or absolute value. Thus a system of comparisons must be established whereby any measurement made can be related to the standard value. Figure 7.1 gives a possible chain of relationships by which everyday measurements in electrical quantities may be evaluated in terms of the accepted standard values.

7.1. STANDARDS

7.1.1 National Standards

Organisations such as the National Physical Laboratory (N.P.L.) of Great Britain, the National Bureaux of Standards (N.B.S.) of the U.S.A. and their equivalents in other countries have expended considerable effort in determining the absolute value of electrical quantities (see appendix I). That is to say quantities such as the ohm, the ampere and the farad, have been determined in terms of the fundamental quantities of mass, length and time with the greatest possible precision although uncertainty still exists in their exact values, even if it is only a few parts in 1 000 000 000.

Such precise determinations of electrical quantities are necessary so that comparisons of the standards used in every country may be made, and enables engineers and scientists throughout the world to have a common set of references: for example, 1 Russian volt = 1 American volt = 1 U.K. volt = 1 Australian volt, etc., and performance of equipment and measurement of physical phenomena are made on a common basis.

In electrical measurements, the standards of greatest importance are those of voltage, resistance, capacitance, inductance and frequency, it being possible to derive other quantities such as current and power from those listed. The absolute determination of electrical quantities is in itself a science[1,2,3] and an appreciation of their necessity should be sufficient at this stage, where the absolute

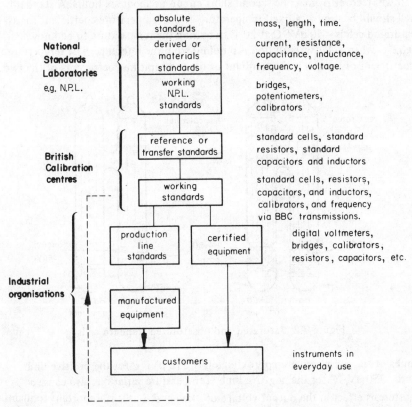

Figure 7.1. Relationship chain between instruments in everyday use and the absolute standards

standards can be considered to be the reference by which derived or 'material' standards are calibrated.

Voltage. Since a material standard of current is not easy to produce, the material standard of voltage in the form of a standard cell is the generally adopted quantity. The two common forms are the Weston saturated cell (which has a terminal voltage of between 1.018 55 V and 1.018 65 V at $20°$ C) and the Weston un-saturated cell (which has a terminal voltage of between 1.0190 V and 1.0194 V over the temperature range $10–40°$ C). Thus the material standard of voltage is a chemical battery in which the positive electrode is mercury and the negative electrode a cadmium–mercury amalgam, with an electrolyte of cadmium sulphate dissolved in dilute sulphuric acid (see figure 7.2). The internal resistance of a standard cell is of the order of 750 Ω; drawing current from it will cause its terminal voltage to fall so any circuit using a standard cell as a reference must have a very high impedance. However, permanent damage to a standard cell is unlikely to occur if current is drawn from it, but the cell will take time to recover. A good cell will recover from a one second short circuit in about six hours. A standard cell should be used at a fixed temperature as the temperature coefficient for a saturated cell is $-40 \, \mu V/^\circ C$ at $20^\circ C$, requiring the temperature to be known to $0.02^\circ C$ when a $1 \, \mu V$ resolution is required from the reference. The temperature coefficient of the cell is the resultant of the temperature coefficients of the two

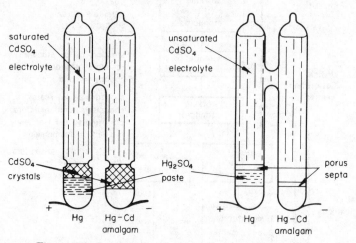

Figure 7.2. Saturated and unsaturated standard cells

limbs of the cell which are approximately $+310 \, \mu V/^\circ C$ for the positive limb and $-350 \, \mu V/^\circ C$ for the negative limb. Temperature variations also cause a hysteresis effect in the output voltage of a cell and thus it is important to maintain a standard cell under constant temperature conditions if possible.

Standard cells are also affected by vibration, and care must be taken when transporting a cell to protect it from extreme vibration, but providing that the cell is in good condition it will recover from the effects of transportation within a few days.

Resistance. The resistance of a resistor is affected by a number of factors, for example, temperature, strain, frequency, contact involving dissimilar metals, and metal purity. In the construction of a standard resistor it is essential that the effects of these factors should be reduced to a minimum and resistors of the highest stability are made from bare annealed wire mounted with as little constraint as possible in a sealed container, which may be oil filled to obtain temperature control. This form of standard is usually limited to resistors having a value of $1 \, \Omega$ or less, as for the wire to be self supporting it must be of moderate dimensions. D.C. resistors of higher values are made from aged covered wire wound on metal formers having the same temperature coefficient of expansion as that of the resistance wire.

Standard resistors are usually of the four terminal kind where the resistance is defined as the ratio of the potential difference across one pair of terminals to the current through the other pair of terminals. If a standard resistor is to be used on a.c. its construction will have to be such that it is unaffected by frequency, that is its inductance and capacitance should ideally be zero. This condition cannot be realised and a number of a.c. resistor forms have been developed which have superior a.c. performance to the simple wound resistor[4]. A simplified analysis of resistor performance when carrying a.c. is given in chapter 4. It should also be noted that the value of a resistor is likely to be different on a.c. from its d.c. resistance (see page 123).

Reactance. Standard capacitors are either gas filled (for use at high voltages) or have mica as a dielectric for lower voltages. Mica is used since it has the best overall insulation performance (see page 127), that is a fairly high permittivity, high electric strength, low power factor, high mechanical strength, low water absorption, chemical inertness, insensitivity to temperature, and can be split into thin uniform sheets. Other materials may be found to be superior in one or two properties but none compares in overall performance. The worst enemy of the standard capacitor is water vapour or moisture since its effects are cumulative causing a change in the permittivity and hence in the capacitance, and thus a standard capacitor should be hermetically sealed. A mica capacitor may be constructed by using either foil as the plate or by depositing silver on the faces of the mica.

An air dielectric capacitor must also be sealed since the permittivity of air changes with humidity, temperature and pressure. A gas filled capacitor usually contains dry nitrogen at a pressure above atmospheric. An additional temperature effect that will be present in both forms of capacitor is that of size variation; consequently, to attain highest precision in measurement of capacitance, temperature conditions must be controlled. To control 'end effects' and stray capacitance a standard capacitor will be fitted with guard rings and mounted in a screening can (see page 143).

Although capacitance is a more usual standard of reactance due to its greater purity and stability, standards of inductance are not uncommon, but since an exact calculation of inductance is not possible, material standards of inductance are calibrated using standard capacitors. Standard inductors are air cored with nonmetallic formers of materials such as marble or ceramic so that the former has a temperature coefficient of expansion similar to that of the winding. They must be either a mutual or self inductor, the former having the advantage of being an active as opposed to a passive standard. Both types are subject to variation in value due to environmental changes and some are constructed on formers having temperature coefficients of opposite sign in orthogonal directions so that the overall change in inductance with temperature is small. Another factor which will affect the value of an inductor is frequency which changes its apparent resistance and the effect on stray capacitance. A simplified analysis of the variation of the properties of an inductor with frequency is given in chapter 4.

Frequency. The standard of frequency is now taken as the frequency of electromagnetic radiation emitted from a transition in the caesium atom (see appendix I).

Comparison between this and the previous standard (Ephemeris time)* gave the frequency of caesium as 9 192 631 770 ± 20 Hz. The caesium standard is superior to the quartz oscillator in that it is unaffected by temperature nor does the caesium oscillator's frequency drift with time. Although the caesium oscillator will not operate continuously for as long as the quartz oscillator, it has become customary to combine quartz crystal and caesium oscillators to form a frequency standard known as an atomic clock. The frequency of the quartz oscillator is regulated from time to time to match that of the caesium oscillator.

International comparisons of the frequency of caesium oscillators made of differing construction[5,10] have been found to be within 2 parts in 10^{10}.

7.1.2 Transfer or Reference Standards

So that the material standards of nations and those of instrument manufacturers within a nation can be compared with one another transfer standards must be used. These standards must be unaffected by transportation and time, that is they must be able to recover from any vibration effects and their aging rate must be negligible and known. To ensure that this stability exists for a transfer standard it is necessary to establish a history of measurement and comparison for each transfer standard. Such a history built up over a number of years would contain records of the standard's values at its 'home' prior to transportation to a standards laboratory, the value measured at the standards laboratory, and the value at its 'home' after it has returned. In addition to these values, readings of temperature, humidity, atmospheric pressure, duration of transportation, and lengths of stabilising times should be recorded for each measurement, not forgetting to note the date, time and place of each measurement. In this manner it is possible to establish the effects of aging on a standard, also the point at which it has reached the end of its life as a transfer standard. The question now arises as to which standards make the best transfer standards. This is, to an extent, governed by the function of a manufacturer and his products, some small manufacturers merely requiring their test equipment to be calibrated periodically at a calibration centre. A major instrument manufacturer would establish its own standards laboratory, or calibration centre having some if not all of the following transfer standards.

Voltage

The commonest transfer standard of voltage is the Weston standard cell. It is also probably the commonest transfer standard of all, but is unfortunately the standard most affected by vibration. It does, however, have a remarkable facility to recover from maltreatment, although since the length of the recovery time is a function of the severity of the maltreatment, every precaution should be taken for the careful handling and transportation of standard cells.

---

* Ephemeris time is the scale according to which the planets appear to move in the most regular manner in their orbits.

Another form of voltage transfer standard that is attaining increasing popularity is the zener diode plus transistor voltage source (see page 203). This has the advantage of being comparatively insensitive to vibration but requires to be mounted in an oven whose temperature is carefully controlled, it being then unaffected by external temperature conditions. This done, voltages stable to 1 or 2 parts per million are obtainable, giving a reference voltage that is used in many models of differential and digital voltmeters, which in turn may be used as a standard of voltage measurement.

Resistance

The four terminal resistor (see page 161) is the accepted transfer standard of resistance and the standards laboratory of an organisation would have a series of resistors, for example 0.01, 0.1, 1.0, 10.0, . . . 100 000.0 Ω, which are sent for periodic measurement to a calibration centre, a history of comparisons between these, the standards in daily use, and the national standards being kept in a similar manner to that adopted for the voltage standards.

Capacitance and Inductance

As with resistance, the transfer standards of inductance and capacitance consist of a series of fixed value standards which are sent for periodic measurement at a calibration centre. Record histories are kept of all measurements.

Frequency

The normal method of obtaining a transfer of the standard frequency from the national standards laboratory to that of an organisation is to use the standard frequency transmissions of an authority such as the BBC which transmits a frequency signal of 200 kHz (1500 m), maintained to within ± 5 parts in 10^9 of 200 000 cycles in 1 s of Ephmeris time, which is in turn defined as 9 192 631 770 cycles of the caesium resonance frequency (see appendix Ib).

Temperature

Since so many of the standard quantities used in electrical measurements are temperature dependent it is necessary for a standards laboratory to have a temperature measuring system related to that of the national standards laboratory. The transfer standard used may either be a good quality mercury in glass thermometer or for greater precision a platinum resistance thermometer (see page 247).

7.1.3 'In House' or Working Standards

In the standards laboratory of a large organisation a set of equipment similar to that used as transfer standards will be maintained as working standards, these being the standards by which checks or calibrations are made on instruments. They are aligned with the national standards by means of the transfer standard, against which they will be periodically checked. The division between working standards and calibration equipment in many standards laboratories will be vague, for example at one time all voltmeters were calibrated against a potentiometer and voltage divider which used a standard cell as a working standard, but the present day tendency is to calibrate all voltmeters (except the highest quality digital voltmeters) with a voltage source which has been calibrated using transfer standards.

One major addition to the working standards list compared with the set of transfer equipment is the ratio standard. For d.c. work this will probably be a self checking Kelvin–Varley divider (see page 153), and for a.c., a ratio transformer (see page 94). Accurate dividers are necessary so that voltages greater than 1.0186 V can be measured. One other addition will be a good a.c. to d.c. transfer instrument enabling a.c. voltage and current measurements to be made in terms of the d.c. voltage standard.

7.1.4 Manufactured Instruments

The dependence of a manufacturer's instruments on the national standards may be derived from one of two main routes (see figure 7.1). Good quality instruments will be individually calibrated with a certificate of calibration relating the performance of the instrument to the standard. The other route of dependence for less accurate instruments will be via a production line standard, which while being an instrument of better accuracy than the equipment it is being used to calibrate, will be inferior to, and calibrated against, the manufacturer's working standards. Instruments related to the national standards by this route will not be sold with a calibration certificate but will be within a specified tolerance of the absolute value.

7.2 ERROR ANALYSIS

From the above section on standards it must be apparent that as the route from an absolute measurement to field measurement is followed the tolerance on a quantity increases. Since field measurements are those of practical engineering, an engineer must be capable of calculating the tolerance on each of the measurements he makes. The errors which can occur during a measurement can be broadly divided into two types: (a) those dependent on the accuracy of the equipment used and known as systematic errors, and (b) those which occur in a haphazard manner and are outside the control of the person performing the measurement, known as random errors.

7.2.1 Systematic Errors

There are a number of forms of error which can be included under this heading, since provided sufficient information is available and has been recorded the total systematic error can be computed from the following types of error.

Constructional errors

This term is used to describe the errors of an instrument which are inherrent from its manufacture or calibration. They are maximum tolerances within which quantities measured by it should lie. For example, the absolute value of a voltage measured at the fiducial value of a 10 V class 0.5 voltmeter will be between 9.95 and 10.05 V. This form of error underlines the importance of periodic calibration checks on all instruments for without such measurements an instrument may have deviated outside its specified tolerance.

When more than one error has to be taken into account the effect is additive. For example, suppose the above voltmeter and a 1.0 A class 0.5 ammeter are used to measure a resistance by the Ohm's law method, the meter readings being 10.0 V and 1.0 A, respectively, then the resistance would be 10.0 $\Omega \pm 1.0$ per cent since V could be 9.95 or 10.05 V and I, 0.995 or 1.005 A. Taking the worst possible cases the resistance ($R = V/I$) is

$$\frac{10.05}{0.995} = 10.05\,(1 + 0.005) = 10.1 \text{ that is} + 1.0 \text{ per cent}$$

$$\text{or}\quad \frac{9.95}{1.005} = 9.95\,(1 - 0.005) = 9.90 \text{ that is} - 1.0 \text{ per cent}$$

Thus it should be apparent that the more instruments involved in the performance of a measurement the greater the uncertainty that is likely to be in the value of the measured quantity and the totalising of construction/calibration errors is of considerable importance since, in most cases, these will account for the major part of the total error in a measurement. The above example indicates that when the result is derived from a quotient, the tolerance of the quantities involved is added. The determination of a quantity may, in general, be represented mathematically as a product or a sum of a number of quantities, which are illustrated by the following

(a) As a product:
$$X = \frac{ABC}{P^2 Q^2} \tag{6.1}$$

Taking natural logarithms of both sides gives:

$$\log_e X = \log_e A + \log_e B + \log_e C - 2\log_e P - 3\log_e Q$$

now
$$\frac{\delta(\log_e X)}{\delta X} = \frac{1}{X} \quad \text{or} \quad \delta(\log_e X) = \frac{\delta X}{X}$$

therefore by obtaining derivatives for both sides,

$$\frac{\delta X}{X} = \frac{\delta A}{A} + \frac{\delta B}{B} + \frac{\delta C}{C} - \frac{2\delta P}{P} - \frac{3\delta Q}{Q} \tag{6.2}$$

But in practice the maximum possible error in X is the quantity it is desired to ascertain, and this will only be obtained if the moduli of the terms are used. Thus the maximum error in X would be:

$$\frac{\delta X}{X} = \pm \left\{ \frac{\delta A}{A} + \frac{\delta B}{B} + \frac{\delta C}{C} + \frac{2\delta P}{P} + \frac{3\delta Q}{Q} \right\} \times 100 \text{ per cent} \qquad (6.3)$$

(b) As a sum: $x = y + z + p$

The error in this case is obtained as follows:

The maximum value of x is, $x + \delta x = y + \delta y + z + \delta z + p + \delta p$
The minimum value of x is, $x - \delta x = y - \delta y + z - \delta z + p - \delta p$
Thus the error in x is $\pm \delta x = \pm (\delta y + \delta z + \delta p)$

or $\qquad \frac{\delta x}{x} = \pm \frac{(\delta y + \delta z + \delta p)}{x} \times 100 \text{ per cent} \qquad (6.4)$

and if the errors of y, z and p had been given as percentages,

$$\frac{\delta x}{x} = \pm \left[\frac{y\delta y}{xy} + \frac{z\delta z}{xz} + \frac{p\delta p}{xp} \right] \text{ per cent} \qquad (6.5)$$

An example of this form of error analysis is in summing the errors for the decades of a resistance box. Consider a four decade resistor box having

decade 'a' of 10 x 1000 ± 1 Ω (0.1 per cent)
decade 'b' of 10 x 100 ± 0.1 Ω (0.1 per cent)
decade 'c' of 10 x 10 ± 0.05 Ω (0.5 per cent)
decade 'd' of 10 x 1 ± 0.01 Ω (1.0 per cent)

and set to 5643 Ω.

The error of the set value is:

$$\frac{\delta x}{x} = \pm \left(\frac{5 + 0.6 + 0.2 + 0.03}{5643} \right) \times 100 \text{ per cent}$$

$$= \pm \frac{5.83}{5643} \times 100 \text{ per cent}$$

$$= \pm 0.103 \text{ per cent}$$

Alternatively,

$$\frac{\delta x}{x} = \pm \left\{ \frac{5000}{5643} \times 0.1 + \frac{600}{5643} \times 0.1 + \frac{40}{5643} \times 0.05 + \frac{3 \times 1}{5643} \right\} \text{ per cent}$$

$$= \pm \left(\frac{583}{5643} \right) \text{ per cent}$$

$$= 0.103 \text{ per cent}$$

Thus $x = 5643 \ \Omega \pm 0.103$ per cent   or $x = 5643 \pm 6 \Omega$.

Determination error

Whenever a measurement is made, there will be an uncertainty in the value read from the display, that is, the unknown will lie in a band limited by either the resolution or the sensitivity of the instrument. An example of this form of error is in the use of a Wheatstone bridge, where variation over three positions of the switch connected to the smallest decade of the known arm, fails to produce any change in the galvanometer indication, the reading is taken as the centre one of the three but with an uncertainty of ± 1 in the smallest decade value. Similarly, an uncertaintly will exist in the reading of a pointer instrument where the pointer position will be subject to a certain amount of operator judgment, for example if the pointer instruments referred to in the above section have scales with 100 divisions and the pointer position may be estimated to 0.2 of a division with an uncertainty of ± 0.1 of a division, a voltage read as 8.84 V would have a determination error of ± 0.01 V, a calibration error of ± 0.05 V, (0.5 per cent of the fiducial value) and a combined error of ± 0.68 per cent of reading at 8.84 V.

Approximation errors

When calibrations have to be performed on readings to determine the value of an unknown quantity, simplifed formulae are often used resulting in approximation errors. An example of this is the expression used for obtaining the magnitude of a low resistance using a Kelvin double bridge (see page 89), where $X = SQ/M$ is a simplification of the expression

$$X = \frac{SQ}{M} + \frac{m \cdot r}{r + q + m}\left(\frac{Q}{M} - \frac{q}{m}\right)$$

and the value of the deleted part of the expression should be added to the total error when determining the magnitude of a low resistor by this method.

Calculation errors

In addition to the approximation errors in result determination, another possible form of error encountered lies in the calculation itself. This may be due to the use of a slide rule where there is a determination error similar to that obtained in reading a pointer instrument, and an allowance of 0.1 per cent should be made for each setting on a 25 cm slide rule. Another form of calculation error results from using tables with too few figures, and this results in rounding errors.

    for example 14.652 rounded to 3 significant figures = 14.7
                14.648 rounded to 3 significant figures = 14.6
                14.65   rounded to 3 significant figures = 14.6 or 14.7
                depending on the operator.

Variations of environment

Pressure, humidity, and temperature can all have an effect on the values of some measured quantities, and corrections for them must be made. For example, the temperature coefficient for a saturated Weston cell is $- 40\ \mu V/^{\circ}C$ of its terminal voltages if this is $1.018590\ V \pm 5\ \mu V$ at $20^{\circ}C$, at $21.2^{\circ}C \pm 0.05^{\circ}C$ its terminal voltage will be $1.018542\ V \pm 7\ \mu V$, the increase in uncertainty being due to the tolerance on the temperature. If the environmental effects cannot be allowed for they must be considered under the category of random errors.

Errors due to strays and residuals

The effect of strays are usually most apparent at high frequencies (see also page 122), where for example, the impedance of stray capacitances becomes low and may shunt the measuring instrument. The category also includes: (a) surface leakage effects, which although these are mostly a high voltage problem, they may be present at low voltages, for example a 1 MΩ leakage path between the terminals of a standard cell would mean a 1 μA current drain and, if the cell's internal resistance were 700 Ω, a reduction in its terminal voltage of 700 μV, (b) contact and lead resistance effects, which may be large in low impedance measurements; (c) inductance of connections, giving a high series impedance at high frequency, and (d) thermoelectric effects at junctions between dissimilar metals, which are corrected for by averaging readings taken with the polarity of the supply reversed.

Aging of equipment

This form of error is determined from the history of the equipment, that is the periodic calibration checks which give the trend of variations with time, and if it is too large to be neglected, the equipment should either be replaced or recalibrated.

7.2.2 Summation of Systematic Errors and Confindence Levels

All the above types of error should be considered in order to obtain the total systematic error[6,7]. They should be added so that the resulting tolerance is the maximum within which the absolute value of the measured quantity will lie.

In practice, however, it is unlikely that all the errors involved in a measurement will be additive, that is the 'probable error' in the measurement will be less than the total error arrived at by a summation which gives the maximum error. This probable error, when expressed as a tolerance, is often termed a 'confidence level' and is arrived at either by deduction from a series of comparison measurements, or by calculating the root of the sum of the square values of the main error terms.

Example. (a) Suppose the value of a nonreactive resistor measured using a transformer bridge was 1766 Ω and a measurement of the same resistor using a Wheatstone bridge was 1764 Ω. If the errors of both bridges are specified as being less than $\pm\ 0.1$ per cent, then from the first method the value of the resistor is between 1764.2 and 1767.8 Ω, and from the second its value is between 1762.2 and 1765.8 Ω; if both bridges are performing within their specifications, the true value of the resistor must be between 1764.2 and 1765.8 Ω, that is, 1765 ± 0.8 Ω or 1765 Ω ± 0.05 per cent.

Example. (b) The 'probable error' in the measurement of the 10 Ω resistor using an ammeter and voltmeter in section 7.2.1 (*constructional errors*) would be

$$\frac{\mathrm{d}R}{R} = \pm \left(\frac{(\mathrm{d}I)^2}{(I)} + \frac{(\mathrm{d}V)^2}{(V)} \right)^{\frac{1}{2}} = \pm (0.5^2 + 0.5^2)^{\frac{1}{2}}$$

$$= \pm 0.7 \text{ per cent}$$

as opposed to the total maximum error of 1.0 per cent.

It must be emphasised that the probable error should only replace a total error when experience has shown it to be justifiable and when the operator is confident that the total error used in the determination of the probable error was the maximum error possible. In addition to this, it should be clearly marked if the result of a measurement is quoted with a probable error, rather than its total error. It is preferable to give both the probable and maximum error values.

7.2.3 Accidental or Random Errors

The errors so far considered have all been calculable, but external influences beyond the control of the operator may have an effect on a measurement. Such quantities are; variations in the supply voltage, mechanical vibration, changes in humidity and pressure, noise (electrical and acoustic), Brownian movement, etc. These phenomena, which are unrelated to the measurement in progress, give rise to errors which are purely random in nature. Whilst they cannot be eliminated their effect can be reduced statistically by taking a large number of readings, and determining the mean or average value which is likely to be nearer the conventional true value than any one individual reading.

The scatter of readings about the mean value gives a measure of the amount of random error involved in a measurement. Ideally this should be small, but the occasional reading, which is very different from the mean, occurs. This rogue value should not be ignored since it may be a true value at that instant in time and result from some hidden systematic error, which has changed during the course of the measurement. If the results of a measurement are subject to random errors, then as the number of readings increases they should approximate to a Gaussian or Normal distribution which can be checked by plotting a histogram (see figure 7.3) that is a graph of the number of occurrences against the value of reading, and establishing that the median (centre line) of the curve coincides with the mean value.

To estimate the probable random error for a set of readings[7], it is necessary to deduce the magnitudes of the observation values within which half the readings lie. Let these values be $+ dx$ and $- dx$ for observations which have a mean value of $\bar{x}$, then the probability of a reading lying within $\pm dx$ of $\bar{x}$ is 50 per cent and the probable random error may be quoted as $\pm dx$.

A more precise method of evaluating the randomness of a set of observations is to calculate the standard deviation for the distribution. Since the sum of the

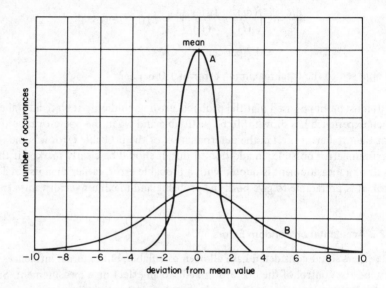

Figure 7.3. Histograms with normal distributions

deviations for all the points in a distribution will be zero, the standard deviation for a set of observations is obtained by calculating the square root of the mean of the sum of the squared deviations that is

$$\sigma = \left[\frac{(x - \bar{x})^2}{N} \right]^{\frac{1}{2}}$$

where $\bar{x}$ is the mean value,

x is an individual value,

and N is the number of values.

For a normal distribution the chance of a valid point lying outside $\pm 1.96\sigma$ is 5 per cent and outside $\pm 3.09\sigma$ is 0.2 per cent. Also for a normal distribution the probable random error is 0.6745σ. The magnitude of σ is a clear indication of the quality of the distribution; in figure 7.3 the curve A would have a σ of 1 while for curve B $\sigma = 3$.

7.3 CALIBRATION PROCEDURES

Environmental conditions will affect the performance of many electrical instruments; therefore calibrations must be conducted under known conditions of temperature, pressure, and humidity. It is therefore important that a room in which temperature, and preferably humidity as well, are controlled, be set aside solely for the calibration of instruments. The requirements for approval of a laboratory under the British Calibration Service[8] (B.C.S.) scheme are for the temperature to be maintained at $20°C \pm 2°C$ (or $23°C \pm 2°C$) and the relative humidity to be between the limits of 35 per cent and 70 per cent. The effects of variations in atmospheric pressure on the performance of electrical instruments are in general small, except in the measurement of high voltages by sphere gap breakdown where they are of extreme importance and must be allowed for. However, the facility to measure atmospheric pressure in a calibration room for electrical instruments should not be overlooked as it may constitute an undetected systematic error in a particular measurement. For example, the dielectric constant of air is pressure dependent.

It has been shown in section 7.1 that a hierarchy of reference exists between the instruments used in day to day engineering measurements and the national or absolute standards. Thus two alternative methods of calibration are available; one in which the instrument to be calibrated is compared with an instrument of superior accuracy or, two, where the instrument under calibration is supplied with its measuring quantity in the form of a known reference. In either method of calibration, the calibrating equipment should have a resolution and error of measurement at least one order better than the instrument to be calibrated, so that the total error in the calibrating equipment is less than or equal to the determination error in the instrument being calibrated. B.C.S. stipulate that the uncertainty in the standard should be less than a quarter of the uncertainty in the instrument under calibration.

7.3.1 Comparison Methods

D.C. Analog (pointer) instruments

Many of these instruments can be inexpensively (in terms of capital equipment) calibrated by using higher grade instruments, for example, a set of Class 0.2 moving coil instruments can be used for calibrating similar instruments whose tolerance is 1 per cent or more, it being reasoned that the calibrating instrument should be used at, or near, full scale deflection. To calibrate the set of Class 0.2 instruments a d.c. potentiometer, digital voltmeter or differential voltmeter whose error is less than 0.01 per cent can be used with the appropriate shunts and voltage dividers. Calibration checks on pointer instruments are normally made at fiducial value and 70 per cent of this value, these calibration checks being performed once a year.

A.C. Analog (pointer) instruments

Low grade, low frequency (20–250 Hz) pointer instruments can conveniently be checked against precision electrodynamic instruments (see page 15), which can in turn be calibrated against a true r.m.s. sensing differential voltmeter or a d.c. standard (differential voltmeter or potentiometer). The waveform of the supply used for calibration purposes must have a very small distortion (better than 0.5 per cent) and be of known frequency.

To calibrate pointer instruments for use in the audio frequencies, a true r.m.s. sensing digital or differential voltmeter (see page 188) may be used, alternatively for low error and high frequency calibrations the combination of a good quality a.c. to d.c. thermal converter and d.c. potentiometer could be used, the scaling of a.c. voltages being performed by an inductive divider (see pages 94 and 156).

Analog (graphical) instruments

Pen recorders

Most potentiometric recorders have an internal reference standard which is used to standardise the pen recorder. External calibration can quite simply be affected by using a d.c. potentiometer. To check the calibration of a moving coil pen recorder, a good quality moving coil instrument of suitable sensitivity may be used.

Light spot recorders

Since the specification accuracy of most recorder galvanometers is ± 5 per cent and in general they are used in conjunction with matching and scaling circuits (see page 59) made up from resistors with ± 10 per cent tolerance, the calibration of the galvanometer and its associated circuit is an essential part of using such a recorder. The method of calibration will largely depend on the application of a particular recorder circuit, for example for d.c. and very low frequency traces, measurements of trace amplitude for various levels of d.c. input (measured with a moving coil instrument) will suffice. But for higher frequencies, calibrations of trace amplitude at the expected frequency of operation must be made using suitable calibrating instruments.

Cathode ray oscilloscopes

The c.r.o. is not an instrument for making precise measurements, but it is an extremely versatile one and its usefulness can be enhanced if it is properly maintained and calibrated. There are many variations and features available in c.r.os and any calibrations performed on them should be made by suitably qualified personnel guided by the handbook of the particular c.r.o. However, the measuring properties of commonest interest to the c.r.o. user are the timebase and trace amplitude accuracies, and these can, on most oscilloscopes, be checked against the internal 'cal' source; which is usually a square waveform of stated amplitude and frequency (see page 70). The other parameter of c.r.o. performance which should be regularly checked (when fitted) is the balance between the inputs of a differential amplifier.

Bridges and comparison methods of measurement

Most instruments of this kind consist of two parts, the ratio arms and the known variable arm. To check the accuracy of one ratio against another, a circuit of the form in figure 7.4 may be used, where zero voltage difference is first obtained between points aa by adjusting the lead compensator LC1 and then between cc

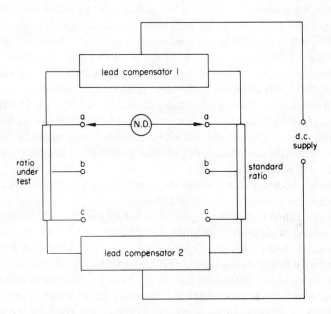

Figure 7.4. Circuit for checking resistance ratios

by adjusting LC2 (these connections being cycled until no further adjustments are required to either LC1 or LC2 for zero voltage between aa and cc). The ratio under investigation can then be checked against the standard ratio which may suitably be a Kelvin–Varley divider, of good quality and preferably selfchecking[9] Another method of checking ratios, such as those used for Wheatstone bridge ratio arms, necessitates the use of a 'build up' resistance box[1] containing $10 \times 10\ k\Omega$ resistors each having an individual tolerance of ± 0.01 per cent and capable of being connected in series or parallel to give resistance values of $1\ k\Omega$, $10\ k\Omega$, and $100\ k\Omega$. Consider the build up box in its $1\ k\Omega$ condition, connected to the X terminals of a Wheatsone bridge (see page 88). Then the 1000/1000, 100/100, 10/10, and 1/1 ratios can all be checked one against the other using the measuring arm. The 10/1 and 100/1 ratio combinations are checked by using the decades of the measuring arm and counterchecked using the build up box.

To check the calibration of the decades of the variable arm of a bridge it is necessary to use a good quality decade resistance box, capable of precise adjustment, and having a stability of better than 1 p.p.m. over the period of each measurement. The process involves first checking the selfconsistency of the variable arm, and then the accuracy of it.

To establish the selfconsistency of the variable arm, use is made of the fact that an integer value may be represented by more than one bridge setting, for example 1 kΩ, may be expressed as either 1000 Ω or 099$\overline{X}$ Ω, the $\overline{X}$ denoting a 10 setting of the smallest decade. The procedure is then to set the variable arm to 1005, the ratio arms to 1000/1000, and balance the bridge by varying the resistance box connected to the measuring terminals, imperfections in the balance being interpolated from the galvanometer deflections for a unit change in the last digit of the variable arm. The bridge is then rebalanced, without changing the value of the resistance box, using the other decades of the variable arm, that is 09X5 plus an interpolation value from the galvanometer. This process is repeated for each integer of each decade, the inconsistency of each bridge setting being found by subtraction. For example, if balances were obtained for the 1000 value as 1005(3) and 09X4(6) the inconsistency would be 700 p.p.m. or 0.07 per cent. This procedure is described in greater detail in a paper by P. M. Clifford in reference 1.

The absolute accuracy of the bridge can subsequently be evaluated by using it to 'measure' the values of two standard resistors.

The above method while referring to a Wheatstone bridge, may be adopted as a general method of calibration for any form of bridge, and as such is a fairly simple means of checking bridges. Other methods are used[2] but are more complex and better suited to calibrating bridges that have the smallest of errors.

Another form of null instrument used in day to day measurement is the d.c. potentiometer, and since this consists of a resistive voltage divider its self-consistency can be checked by an adaptation of the decade checking process described for use with bridges. The 'unknown' in this case is derived from the voltage division, by a Kelvin–Varley divider, of a stable voltage source[1].

Digital instruments

Differential voltmeters

The errors of these instruments are dependent on two principal features: one, the stability of the internal voltage reference; and two, the dividing accuracy of the divider network. The first of these may be checked by determining the drift rate of the internal voltage reference with respect to a constant environment standard cell, and the second feature by procedures outlined above for determining errors in ratio networks.

Counter/Frequency

These may be checked against standard oscillators or standard frequency trans-missions from broadcasting stations by straightforward measurement and adjustment, or, for precise calibration, by comparison over a period of time[10].

Digital Voltmeter

The method adopted to calibrate a digital voltmeter will be dependent on the quality of its measurement capabilities, that is one grade of d. vm. may be calibrated by the next higher grade, or by using a suitable differential voltmeter. To calibrate the best quality d. vm., a stable voltage source, a precision voltage divider, and a precision d.c. potentiometer would have to be used[11]. This latter method could, of course, be used to calibrate any d. vm.

7.3.2 Calibrators

Variations of the methods outlined in 7.3.1 are, in many cases, those in general use; and are satisfactory procedures except that they require a stable voltage or current source, a highly trained operator, are time consuming and consequently expensive. Since by far the majority of instruments in daily use are of the lower grades of accuracy, it is desirable that the methods used for the calibration of these instruments should be inexpensive. To meet this requirement the logical step is to devise a calibration source whose output (voltage or current) may be set to a desired level, within a known tolerance, thus eliminating the calibrating instrument from the system. Therefore the essential requirements for a calibrator are:

(a)   a high degree of stability,
(b)   the waveform should be 'pure' on a.c. and ripple free on d.c.,
(c)   an output, unaffected by, and having errors at least one order less than, the instrument undergoing calibration, and
(d)   an ergonomic design for efficient operation.

A range of calibrators (voltage and current) are marketed, with graded accuracies, for d.c. only, a.c. only (fixed or variable frequency), and as combined units. An additional feature sometimes incorporated is the use of remote and/or programmed control. Broadly speaking they may be divided into two types (a) meter calibrators and (b) calibration systems.

Meter calibrators

The operation of these devices is a development from that of a regulated power supply (figure 7.5), where the output voltage level is controlled by a resistive ratio ($V_0 = - V_{ref}(R_2/R_1)$), and the reference voltage, which may be a zener diode circuit. The accuracy requirement for a meter calibrator will in general be that for a precision instrument, that is ± 0.2 per cent.

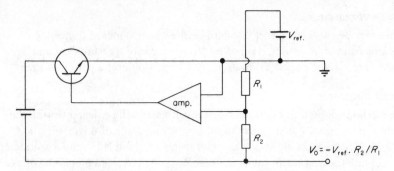

Figure 7.5. Outline diagram of a regulated power supply

The ranges for one commercial calibrator being;

(a)   D.C., and a.c. (50 Hz) voltage
      0.01–1 V; 0.1–10 V; 1–100 V; 10–1000 V.
(b)   A.C. and d.c. current
      1–100 μA; 0.01–1 mA; 0.1–10 mA; 1–100 mA; 0.01–1 A; 0.1–10 A.
Errors on a.c. are ± 0.4 per cent of setting 0.1 per cent of range, and on d.c.
      is ± 0.2 per cent of setting ± 0.1 per cent of range.

Some meter calibrators incorporate a percentage deviation indicator which
enables rapid evaluation of instrument errors.

Calibration systems

These are really sophisticated versions of the meter calibrators mentioned above,
they tend to be designed either for a.c. or d.c. use and have accuracy specifications
better than the best measuring instruments, that is they may be used to calibrate any
instrument from a class 1 pointer instrument to a 0.001 per cent d. vm.
      The heart of such a system is the stable d.c. voltage source which must have
defined levels capable of being set with a high degree of precision (for example
0.003 per cent of setting, 0.1 p.p.m. resolution of range). Such a d.c. source will
not retain its calibration indefinitely and provision must therefore be incorporated
to facilitate the periodic calibration of the voltage source, against a sensitive null
detector (1 μV f.s.d.). The versatility of the system may be increased by the
addition of a selfcalibrating Kelvin–Varley divider[9], the linearity of which can be
made better than 0.1 p.p.m. and can then be used to check the reference divider.
Such a d.c. voltage and ratio calibration system can be used for:

(a)   calibration of power supplies – uncertainty 5–7 p.p.m. plus the
      standard cell accuracy
(b)   calibration of voltmeters – uncertainty 5–8 p.p.m.
(c)   ratio calibration – uncertainty in linearity 0.1 p.p.m. of input

X Y Z. Co. Ltd STANDARDS LABORATORY CALIBRATION REPORT		Date Ref. Sheet of	
Instrument type Manufacturer Ranges		Serial No. Previous calibration Date Ref.	
Method of calibration			
Range	Instrument reading	Calibrating value	% Error
Remarks			
Temperature	Humidity		Pressure
Copies to:		Calibrated by: Date	

Figure 7.6. Example of a record form for the calibration of an instrument

(d)   calibration of voltmeters – uncertainty of 20 p.p.m. (using the d.c. voltage source without the reference divider and standard cell)

(e)   measurement of d.c. voltages as a differential voltmeter with an uncertainty of 5–8 p.p.m.

If a high quality a.c. to d.c. transfer instrument is incorporated in the system it may be used for precision a.c. measurements.

An a.c. calibration system will have the same basic form as the d.c. one, a stable calibrated voltage source, a precision divider (inductive) and a sensitive null detector. The voltage source must have a low harmonic content in its output, good stability, and preferably a wide frequency range, for example, 10 Hz–100 kHz. To establish the errors of the calibrations system a good quality a.c. to d.c. transfer instrument must be incorporated so that comparisons with a standard cell may be performed.

It should be appreciated that the errors of an a.c. calibrator will be greater than those of a d.c. one. The least errors available for a.c. calibration systems being around 0.01 per cent, whereas 0.001 per cent is not uncommon for a d.c. system[12]. Some systems may be programmed[13] and this leads to an even further reduction in calibration time.

7.3.3 Records

The importance of calibrating records cannot be over emphasised. It has been stated above that the calibration history of a transfer standard must be built up from the records of periodic calibration checks, and this applies to any instrument; that is deterioration with age can only be detected from trends in the record of its calibration checks. It therefore becomes apparent that the record of every calibration of an instrument should contain full details of the equipment used, the environmental conditions, the circuit arrangement, and the procedure adopted for the evaluation of the results. All the readings taken and results obtained should be clearly tabulated so that any error trends are conspicuous. Figure 7.6 indicates the type of forms which are used by standards laboratories[1,8]. A procedure must also be instigated by which instruments have their calibration checked at regular intervals, and perhaps the simplest method of doing this is to label each instrument with the date of the last calibration check and the date when the next calibration check is due. It is then the responsibility of the operator to ensure that he is not using an instrument after its calibration check is due.

REFERENCES

1   J. R. Thompson (Editor). *Precision Electrical Measurements in Industry*, Butterworths, London (1965)

2   D. S. Luppold. *Precision d.c. Measurements and Standards*, Addison Wesley, Reading, Mass (1969)

3   H. Buckingham and E. M. Price. *Principles of Electrical Measurements,*
     English Universities Press, London (1955)
4   F. E. Terman and J. M. Pettit. *Electronic Measurements,* McGraw-Hill,
     New York (1952)
5   'Flying Clock Experiment'. *Hewlett Packard Journal,* (Dec. 1967)
6   D. Karo. *Electrical Measurements, 2nd ed., Pt 1,* Macdonald, London (1953)
7   J. Topping. *Errors of Observation and Their Treatment,* Chapman and
     Hall, London (1971)
8   *British Calibration Service.* General Criteria for Laboratory Approval (1967)
9   Jan Sliper. 'Advanced d.c. Calibration Techniques'. *E.E.E. Eruopean Seminar*
     (April 1969)
10   'Frequency and Time Standards'. *Hewlett Packard Application Note 52*
     (1965)
11   'Which D.C. Voltmeter'. *Hewlett Packard Application Note 69* (1965)
12   *Hewlett Packard Journal* (April 1970)
13   E. R. Aikin and J. L. Minck. 'Automating the Calibration Laboratory'.
     Hewlett Packard Journal, **24,** No. 8 (April 1973)

8

Transducers

Definition

A transducer is a device or element that may be used to convert an input signal into an output signal of a different form[6]. It may therefore be a device that converts a mechanical variable into an electrical one, for example a tacho-generator, or it may be capable of converting an electrical signal into a mechanical one, for example a galvanometer. Commercial usage of the term is generally accepted as meaning a device that converts a physical phenomenon into an electrical signal.

In many cases the conversion may be via an intermediate stage, that is, the measurand (for example pressure) is first converted to a mechanical displacement which in turn is converted to an electrical signal. The mechanical conversion is accomplished by one of two fundamentally different methods, which are:

Fixed reference devices. In which one part of the transducer is attached to a reference surface, and the other part connected, either directly or via a linking mechanism, to the variable as in figure 8.1. Should the displacement be small, some means of mechanical magnification or electrical amplification must be used to obtain a satisfactory sensitivity.

Mass-spring or seismic device. In these transducers there is only one contact or anchor point, this being the attachment of the transducer base to the point where the variation is to be measured. The motion of the measurand being inferred from the relative motion (δ) of the mass (m) to that of the case (figure 8.2), and will depend on the size of the mass and the stiffness of the spring (k), the magnitude of the damping being determined by the damper (C). The seismic type of transducer is indispensable in the study of movements and vibrations in any form of vehicle, it has a sensitivity—frequency characteristic similar to a recorder galvanometer (see page 53), and it may be shown that to avoid amplitude distortion when using a seismic displacement transducer ω (the frequency of sinusoidal movement) must be greater than $2\omega_0$ (where ω_0 is the internal

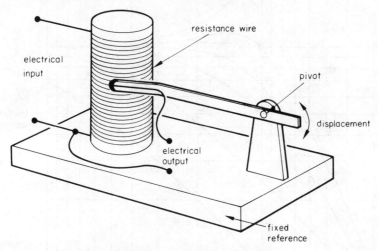

Figure 8.1. Fixed reference transducer

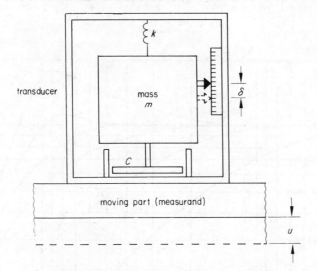

Figure 8.2. Seismic or mass spring transducer

resonant frequency of the transducer), while for an accelerometer undergoing sinusoidal acceleration, the ratio ω/ω_0 must be less than 0.2. In figure 8.3 the effects of damping magnitude on both of these characteristics are shown[1,2].

The relationships between sinusoidal displacements, velocities, and accelerations are shown in figure 8.4 from which:

$$\text{peak acceleration} = \omega^2 . S_{max} \quad \text{m/s}^2$$

$$= \frac{\omega^2 . S_{max}}{9} \quad \text{in gs of acceleration.}$$

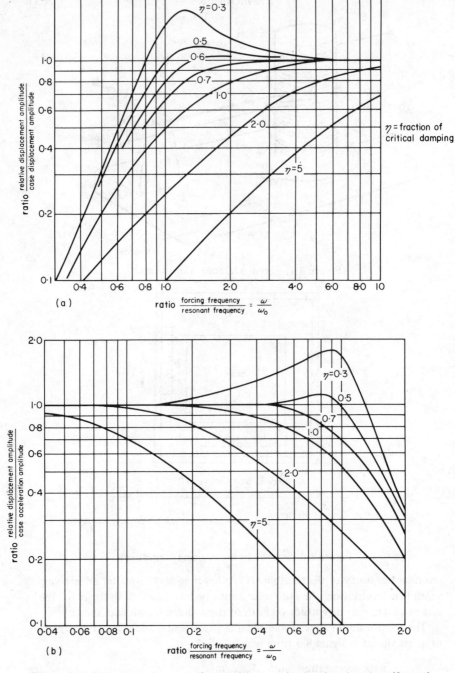

Figure 8.3. Characteristic curves for seismic transducers showing the effect of
damping magnitude on displacement amplitude of internal mass.
(a) Response to sinusoidal displacement of transducer case;
(b) response to sinusoidal acceleration of transducer case

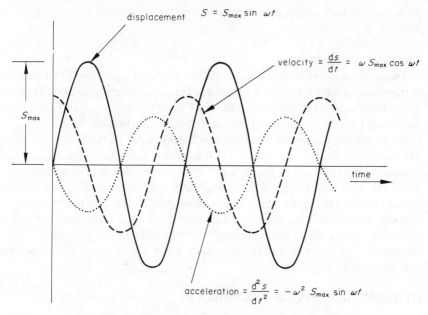

displacement $S = S_{max} \sin \omega t$

velocity $= \dfrac{ds}{dt} = \omega S_{max} \cos \omega t$

S_{max}

time

acceleration $= \dfrac{d^2 s}{dt^2} = -\omega^2 S_{max} \sin \omega t$

Figure 8.4. Relationships of oscillatory motions

Table 8.1. Table of measurands against transduction methods

	Resistance Change	Reactance Change	Electro-magnetic	Semi-conductor	Digital	Thermo-electric
Acceleration	*	*	*	*		
Displacement	*	*	*	*	*	
Flow	*	*	*			
Force	*	*		*		
Humidity	*					
Level	*	*				
Pressure	*	*	*	*		
Temperature	*			*	*	*
Thickness	*				*	
Velocity	*	*	*	*	*	

An accelerometer's output performance usually being specified in terms of peak mV/g.

Transducer classification

To convert a mechanical movement into an electrical signal, one of a number of transduction methods may be employed. Transducers may be classified either by the transduction method employed, or by the function it is capable of measuring, for example acceleration, displacement, etc. In table 8.1 are tabulated the more common measurands and the possible methods of transduction. The method of classification adopted in this book is that of transduction method, it being considered that in a teaching/introduction text this method of classification is the more logical; whereas in an application orientated book on transducers, classification by use would be more appropriate[6]. It should be noted that in many cases the various transduction methods have equal merit, and that each of the various transducer manufacturers having specialised on a particular transduction method will tend to incorporate it in the majority of the transducers that they make.

8.1 RESISTANCE CHANGE TRANSDUCERS

The methods of transduction involving a change of resistance are perhaps the easiest to understand and will therefore be described first. The variable resistance method of transduction is not widely used but serves to illustrate the underlying principles.

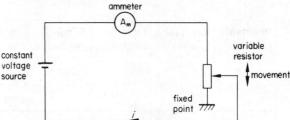

Figure 8.5. Variable resistance position transducer

Consider the diagram in figure 8.5 where the operation of this circuit as a position measuring system simply requires that the sliding contact on the resistor be linked to the motion or displacement under observation. A change in the measurand will cause a change in the resistance in the circuit and a consequent change in the current. If the ammeter were calibrated in suitable units a continuous display of the measurand would thus have been obtained. For such a transducer to operate satisfactorily the voltage source must have a constant magnitude irrespective of circuit resistance, so that the current in the circuit is only dependent on the variations in the resistance R. Another drawback of this type of transducer is that if a plus and minus variation is to be observed, current will be flowing in the circuit when the measurand is in the zero or reference position.

8.1.1 Potentiometric Transducers

The variable resistance transducer results in a current variation which is a function of the magnitude of the measurand. In a potentiometric transducer the output is a voltage variation, that is the resistive divider is used as a potentiometer or potential divider, figure 8.6, and is in general a more satisfactory arrangement.

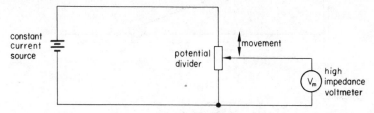

Figure 8.6. Principle of the potentiometric position transducer

A potentiometric transducer should be supplied from a constant current source, the transducer output being monitored by a high impedance instrument so that the loading effects across the potentiometer are negligible. It will be seen that mechanically coupling the measurand to the sliding contact of the potentiometer will result in an output voltage that will be proportional to the displacement of the wiper from one end of the resistive path.

Since the resistance potentiometer may be used to convert a displacement into a proportional electrical signal, the measurement of quantities such as flow, force, pressure, humidity, etc., can be performed by a fixed reference potentiometric sensor when a suitable quantity to displacement converter is positioned between the sliding contact and the measurand. Figure 8.7 illustrates some of the techniques used to convert physical properties to displacements suitable for applying to the sliding contact.

To obtain measurements of velocity or acceleration from a fixed reference potentiometric transducer, requires that the output voltage proportional to displacement is electrically differentiated once to obtain velocity and differentiated twice to produce an electrical signal proportional to acceleration (see page 130). A signal proportional to acceleration can however be obtained directly from a seismic potentiometric transducer (see figure 8.2) providing its mechanical resonant frequency ω_0 is much higher than the frequency of the acceleration (ω), that is $\omega/\omega_0 < 0.2$.

Both the variable resistance and the resistance potentiometer types of transducer have a resistive element which may be wire wound, deposited carbon film, platinum film, or a resistive compound such as conductive plastic. Since the operation of all these transducers depends on the contact between a slider and the resistance element, their life is related to the wear of this contact, and the frequency of operation limited to a few cycles of displacement per second. The typical life expectancy of this type of transducer is around 3×10^6 cycles of

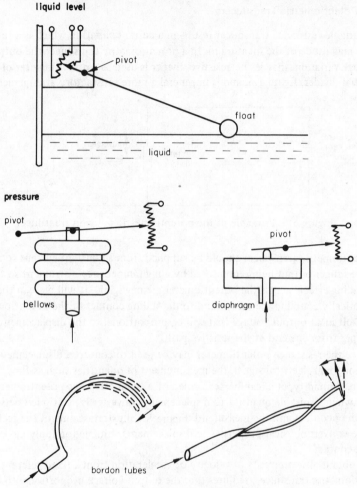

Figure 8.7. Some transducer actuating mechanisms

operation; they are comparatively inexpensive, and the circuitry associated with their use is simple. It should, however, be remembered that if displacement is the quantity being measured, a force has to be provided to overcome the friction of the sliding contact and this may effect the magnitude of the observed displacement.

8.1.2 Resistance Strain Gauges

An important series of transducers is derived from the use of resistance strain gauges, and before considering such transducers the principles and applications of resistance strain gauges should be understood.

If a length of electrical wire is subjected to a tensile force it will stretch, its
length being increased by an amount δl (say), and providing that the elastic limit
of the material is not exceeded the change in length is proportional to the load,
and the wire will revert to its original length when the load is removed. Corresponding to this increase in length there will be a slight decrease in the cross-sectional
area of the wire (the increased length has to come from somewhere), and since
resistance of a conductor = $\rho l/A$, where ρ = resistivity of the material; l = length;
and A = cross-sectional area, the increase in length and the decrease in area will
both contribute to an increase in the resistance of the stretched wire. Obviously
it is inconvenient to have long lengths of resistance wire attached to a test object
so the general arrangement of a strain gauge is one in which the resistance wire
(which typically is 0.025 mm thick) is folded into a grid and mounted on a
backing of paper or bakelite. A development from this is the foil gauge, figure
8.8, which is manufactured by techniques similar to those used in the production
of printed circuits. Such a process readily lends itself to the production of shaped,
special purpose, strain gauges. The size[3,4] of a strain gauge will depend on the
application but resistance gauges are available in a variety of lengths from around
3 mm to 150 mm there being a range of nominal resistance values, preferred
magnitudes of which are 120 and 600 Ω.

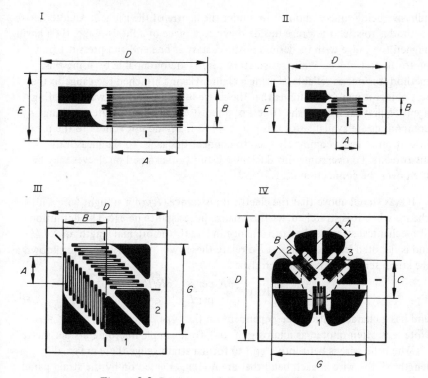

Figure 8.8. Resistance foil gauges (Philips)

Strain gauge attachment

In the majority of strain gauge applications it is necessary to attach the gauge to the test object by means of an adhesive. The thickness of the adhesive layer should be made small to avoid errors in transmitting the size changes in the test object to the strain gauge. This is best performed by using a clamp, curved metal plate, rubber pad, and nonadhesive foil film as indicated in figure 8.9, the excess

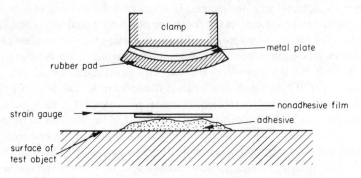

Figure 8.9. Clamping method of attaching strain gauges

adhesive being squeezed out from under the centre of the gauge[5]. An alternative method is to stick the gauge upside down to a piece of adhesive tape, then having aligned the gauge with its desired position start at one end and press it firmly on to the test object. While satisfactory results are obtainable by using this method it is not as reliable as using a clamp. Having attached the gauge to the test object it may be desirable that the gauge is covered with a waterproofing agent to prevent it from the effects of a moist environment. Waterproofing compounds are available for this purpose but many do not adhere to the p.v.c. insulation of leads commonly used to connect the gauge to the measuring instrument. To overcome this difficulty clamps or crimped-on sleeves may be used over the connection leads.

It was stated above that the electrical resistance (R) of a straight wire will change when it is stretched; let this change in resistance be δR. The definition of mechanical strain is the ratio 'change in length/the original length' or $\delta L/L$ and is denoted by the symbol ϵ. To relate these two quantities, a factor known as the strain sensitivity is used such that:

$$\text{strain sensitivity} = \frac{\delta R \text{ per } R}{\delta L \text{ per } L} = \frac{\delta R/R}{\epsilon}$$

and has a characteristic value dependent on the type of resistance wire.
Note. ϵ is often quoted as microstrain ($\mu\epsilon$), for example micrometers per metre.

When the wire is bent into a grid to form a strain gauge there are small lengths of the wire at each bend that are no longer acted on by the strain parallel to the axis of the gauge. This part of the gauge resistance is termed 'dead resistance' and the ratio of dead resistance to total gauge resistance will depend upon the type of fold and length of the gauge.

When a gauge is attached to an object that is stressed parallel to the axis of the strain gauge it may sense two strains: (a) the principal or longitudinal strain which is in line with the gauge axis, and (b) the transverse or Poisson strain which is at right angles to the gauge axis. If the gauge has appreciable dead resistance it will exhibit a transverse strain sensitivity, which is typically 1–2 per cent of the longitudinal sensitivity, this effect may be reduced in foil gauges by increasing the width of material at the bend (see figure 8.8). The output resulting from this Poisson strain acts to oppose that from the principal strain, thus reducing the overall gauge sensitivity.

Gauge factor

The strain sensitivity of a manufactured strain gauge is termed the *gauge factor*, and is less than that of the resistance wire used in its construction, as it will include the effects of the dead resistance, the Poisson ratio of the surface to which the gauge is attached, and the transverse sensitivity. The majority of resistance gauges have a gauge factor of around 2, copper-nickel wires giving values between 1.9 and 2.1; however, iron-chromium-aluminium and iron-nickel-chromium alloys may be used to give a gauge factor of 2.8 to 3.5[6].

Vibratory movement

When a strain gauge is used in a dynamic situation, for example for the determination of strains caused by vibrations or movement, the frequency response of the gauge must be considered. Since a gauge will average the strain over its active length, if the wavelength of the strain variations are of the same order of magnitude as the gauge length, considerable errors will be introduced. In the worst case, if the gauge length is equal to the vibratory wavelength, the effective average strain

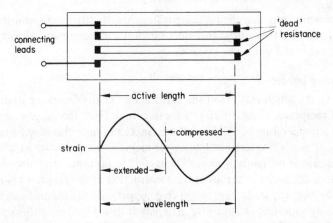

Figure 8.10. 'Worst case' of vibration frequency, occurring when the vibration wavelength equals the active length of the strain gauge

and hence the output, will be zero (see figure 8.10). If the frequency were increased above this point, output would again be obtained but its amplitude could not be easily related to the measurand.

To determine the frequency of vibration at which zero output would occur, requires the application of the expression relating frequency and wavelength to the velocity of sound, that is $f.L = u$. Where f is the frequency of vibration in the test object; L is the active length of the strain gauge; and u is the velocity of sound in the medium to which the strain gauge is bonded. If, as is commonly the case, the gauge is attached to a steel member, in which u is 5×10^3 m/s (approximately), the frequency for zero output is

$$f = \frac{5 \times 10^3}{L} \text{ Hz and if } L = 5 \text{ mm } f = 1 \text{ MHz}$$

alternatively if $L = 150$ mm, $f = 33$ kHz.

These values of frequency are theoretical values, and in practice (due to the effects of bonding) the frequency at which zero output occurs will be considerably less. To obtain an error of 1 per cent or less in the output, the theoretical values should be divided by a factor of 20, giving practical upper frequency limits of 50 kHz and 1.6 kHz for the 5 mm and 150 mm gauges respectively. Since instrument specifications normally quote frequency limits as 3 dB points (where response is 30 per cent down (see page 130)), these may be determined for the strain gauge by $f \approx u/4L$, giving bandwidths B for the above gauges of 250 kHz and 8 kHz respectively. To obtain the bandwidth of the measurement system – when the gauge is joined to a display instrument – the expression for the system bandwidth is:

$$B_{\text{system}} = \frac{B_1 \times B_2}{[(B_1)^2 + (B_2)^2]^{\frac{1}{2}}}$$

is used, remembering that the resulting system bandwidth should be divided by 5 to obtain a system frequency range over which the measurement errors should be of the order of 1 per cent.

Temperature effects

Another factor which may effect the performance of the resistance strain gauge is that of temperature, and its effects are threefold. First, the gauge filament will have a temperature coefficient of resistance, and since this may be as large as 50 p.p.m. per $^\circ$C for some of the copper-nickel alloys (Constantan $\pm$ 20 p.p.m. per $^\circ$C) it cannot be lightly ignored. The second temperature effect results from the presence of dissimilar metals in the construction of the gauge, it being common practice to form the leads from nickel clad copper, and the thermal e.m.f. generated by various resistance wire materials at their junctions with copper varies from about 2 μV/$^\circ$C for Manganin to about 47 μV/$^\circ$C for Constantan. The third temperature effect results from the difference in the temperature coefficients of expansion of the materials of the test object and the resistance wire of the gauge. For example, should the test object expand (thermally) at a greater rate than the gauge wire, the latter will be stretched and to the measuring instrument it will appear as if the test object is being stressed, this phenomenon being known as apparent strain.

To compensate for these temperature effects one of the following methods may be adopted, namely:

(a) the use of selfcompensated gauges. These have a filament material whose temperature coefficient of expansion is approximately equal to that of the test object. Since it is not practical for manufacturers to market gauges with temperature coefficients equal to all the materials that a test object may be made from, the normal compromise is to market gauges with expansion coefficients comparable with the commonly used groups of materials, that is the steel group; the copper alloys; and the aluminum group;

(b) the use of dummy or nonactive gauges in the measuring circuit;

(c) the inclusion of a compensating thermocouple in the measuring circuit or the use of an additional or compensating filament in the construction of the gauge[3].

The methods a and b generally give the most satisfactory results. The measuring circuit used in strain gauge work is usually some form of d.c. Wheatstone bridge circuit. However, when used with strain gauges it is common for it to be operated in the unbalanced condition (see page 112); bridges balance only corresponding to the initial unloaded conditions of the test object, thus enabling a much faster and cheaper recording system than would be possible if the bridge were balanced for each reading. It should be noted that changes in the gauge resistance due to temperature (aR in figure 8.11) may be cancelled out by using arrangements that incorporate a dummy gauge, or alternatively by using four active gauges or two in adjacent arms. The bridge circuit containing an uncompensated gauge has an output voltage that is proportional to the sum of the strain and the thermal effects.

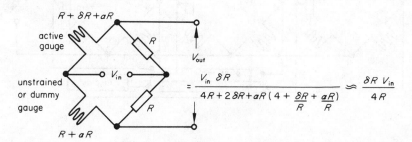

Figure 8.11. Basic strain gauge bridge circuit (see also figures 3.40 and 3.41)

Multipoint measurements

In many applications of the strain gauge, it will be desirable to measure strain at more than one point, for example in determining the effects of a load on a structure.

These necessitate the use of a number of strain gauges which may be arranged so that either they have a bridge circuit and display instrument each or one bridge circuit and display instrument is connected, via a switch, to each strain gauge in turn. The latter method is undoubtedly the least expensive but can introduce problems of its own, for example switch contact resistance, and different zero balances for each gauge circuit[3].

Since it is possible to use one, two, or four active gauges, in determining the strain in one area of a test object, arrangements must be developed whereby each of these situations may be incorporated in a multipoint measuring system. Figure 8.12 illustrates two conditions, the display being an interpretation of the output from an unbalanced bridge, which by suitable scaling would be in terms of microstrain.

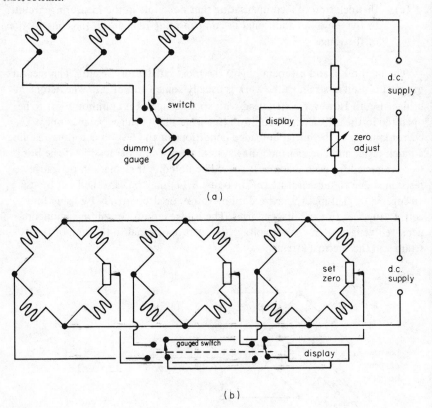

Figure 8.12. Multipoint arrangements. (a) Single active gauge arrangement; (b) four active gauge arrangement

Lead resistance

In many applications the measuring/display instruments will be remote from the active gauges, and the effect of the wires connecting the active gauges to the remainder of the circuit is equivalent to inserting a small resistance in series with it. This resistance will vary due to temperature effects and to reduce its influence additional connections as given in figure 8.13a and b are required and then, the lead resistances are approximately selfcancelling.

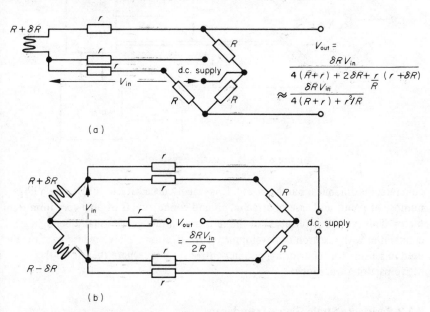

Figure 8.13. Arrangements for selfcompensation of lead resistance. (a) Single active gauge; (b) two active gauges

The above discussion of strain gauge measuring techniques has assumed that all methods use a d.c. supply. While the tendency is towards such methods due to the development of sensitive, drift free d.c. amplifiers, digital voltmeters, and data logging system, the a.c. carrier system has not completely disappared from use. The a.c. system had a considerable advantage in its stability of amplification but stray capacitance and the need to balance its effect from the measuring circuits are a considerable disadvantage. Figure 8.14 illustrates the general principles involved in an a.c. carrier system of measurement.

Applications

The uses of resistance strain gauges are almost limitless but their direct uses may be summarised as applications involving the measurement of stress and strain in existing structures such as aircraft, railway wagons, bridges, cranes, reinforced

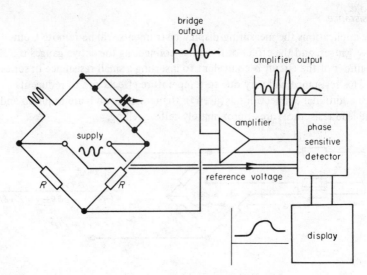

Figure 8.14. A.C. carrier system

concrete, automobiles, buildings, etc. It is usually necessary to investigate a large number of points and ease of attachment and connection is often a very important factor. This type of application of the resistance strain gauge usually being in connection with research or development investigations. They are also extensively used in sensors for monitoring, and in control systems, where they form the active part of a transducer.

8.1.3 Resistance Strain Gauge Transducers

To operate, a strain gauge transducer requires that the phenomenon under investigation shall first be converted into mechanical strain, that is the gauge is attached to an elastic member, within the transducer, which is subjected to a force proportional to the measurand. If the force is small, for example as in the measurement of small pressure differences, an unbonded strain gauge system may be adopted where the strain gauge wire itself acts as the elastic member. Figure 8.15 shows such a sensing element, which may be incorporated in a wide range of pressure transducers simply by varying the area or thickness of the diaphragm that operates the force rod. This type of assembly contains four resistance strain gauges, two being relaxed and two stretched when a force is applied to the spring element. The four gauges are connected to form a Wheatstone bridge, unbalance in the bridge arms in the unloaded condition being compensated for during construction by the addition of small resistances in the appropriate arms. In a good quality unbonded strain gauge transducer consistency of operation is ensured by temperature and pressure cycling during manufacture, the sensing unit being sealed in a container of dry helium, quality checks being

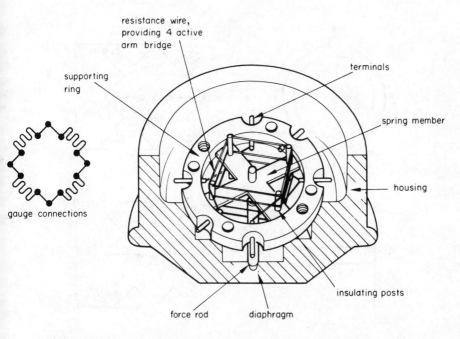

Figure 8.15. Unbonded strain gauge pressure transducer (Courtesy of Bell and Howell)

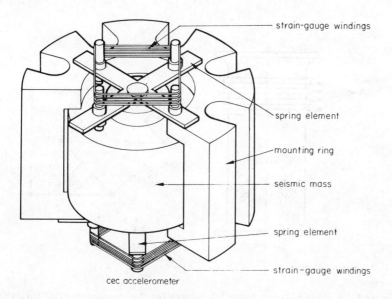

Figure 8.16. Unbonded strain gauge seismic accelerometer (Courtesy of Bell and Howell)

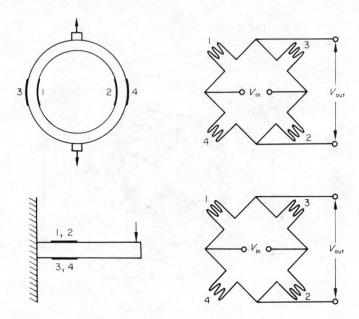

Figure 8.17. Some bonded strain gauge force transducers (load cells)

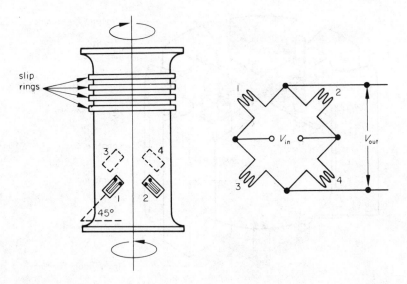

Figure 8.18. Bonded strain gauge transducer for the measurement of torsion in a shaft

conducted at each stage of manufacture of the transducer. The use of unbonded strain gauge transducers is not limited to pressure gauges, figure 8.16 illustrates its application in the construction of a seismic accelerometer. Since many physical phenomena may be converted to a variation of force, a strain gauge element of the type shown in figure 8.15 could be incorporated into transducers to measure such properties as weight, temperature, humidity, flow and viscosity.

The use of strain gauges in the construction of transducers is by no means limited to unbonded gauges[3,4,6], common examples of the use of bonded gauges in transducer construction being in the measurement of force (load cells) figure 8.17, and torsion, figure 8.18.

8.1.4 Other Resistance Change Transducers

Resistance thermometer

Some materials have a high temperature coefficient of resistance (α), and since the resistance (R_T) of such a material, at a temperature T may be related to its resistance at $0°C$ by the expression $R_T \approx R_0 (1 + \alpha T)$ Ω, this property may be made use of in the measurement of temperature. Platinum resistance wire is normally used in the manufacture of resistance thermometers, which may be

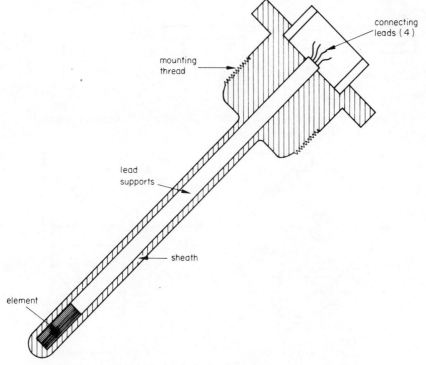

Figure 8.19. Industrial platinum resistance thermometer

formed in a manner similar to a resistance strain gauge[6], alternatively the resistance element may be housed in some form of sheath as shown in figure 8.19.

The resistance thermometer has a high intrinsic accuracy, and they are used as an international standard for comparing temperatures in the range 150—1100 K, and thus their major use is in laboratories where precise temperature measurements are important. However, in some industrial applications, for example where high accuracy may be required, or if signal amplification is undesirable, a platinum resistance thermometer may be used since these transducers exhibit approximately a 39 per cent change in resistance between 0 and $100°C$. They are however fragile, expensive, error producing when mishandled, and have a long $(0.5—10\,s)$ response time (compared with thermocouples).

The variations in thermometer resistance may be measured by: (a) some form of Wheatstone bridge circuit (see figure 8.20) which may operate in a balanced or

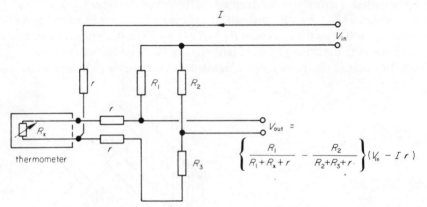

Figure 8.20. Wheatstone bridge circuit for use with three lead resistance
thermometers

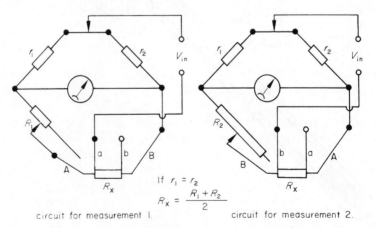

Figure 8.21. Smith bridge — two measurements are made to eliminate lead effects

unbalanced mode; (b) supply from either a constant current or voltage source and recording variations in voltage or current; (c) using a Smith bridge (see figure 8.21); (d) a Müller bridge (see figure 8.22); or (e) an inductively coupled ratio bridge of the type shown in figure 3.20 (see also references 7, 8, 9, 10 and 11). It will be seen that some of these diagrams show the use of three and four terminal connections to the resistance element, the purpose of the multiple

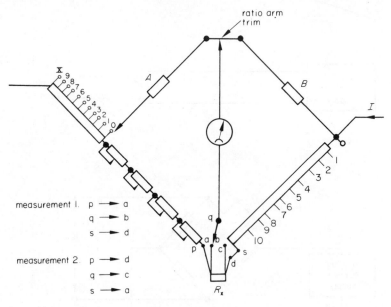

Figure 8.22. Müller bridge

connections being to remove the effects of lead resistance from the measurement. The Smith, Müller, and inductive ratio bridges are all precision devices capable of resolving temperatures to 0.0001 °C and their use is therefore normally limited to the standards laboratory.

Hot wire anemometers

For measuring air flow rate a resistance change transducer consisting of a tungsten or platinum alloy wire through which a current is passed (see figure 8.23) may be used. The magnitude of the current is sufficient to heat the wire, and any movement of air over the wire will cool it and cause a change in resistance, and a consequent change in the voltage drop across the wire. If this volt drop is applied to an oscilloscope or u.v. recorder a display or record of the air flow variations is obtained. The hot wire[4,6] is small (typically 1 mm long, and 0.1 mm dia.) and capable of responding to quite rapid fluctuations in flow. The greatest speed of response is obtained by using the constant temperature arrangement shown in figure 8.23d where a control unit incorporating a feedback amplifier compensates

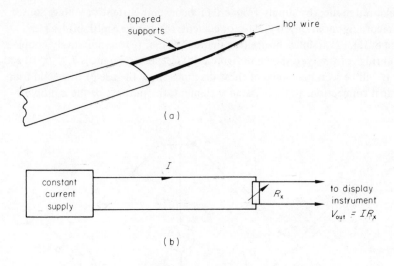

(a)

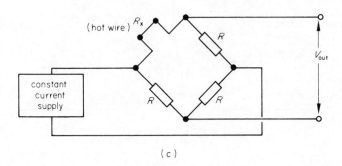

(b)

(c)

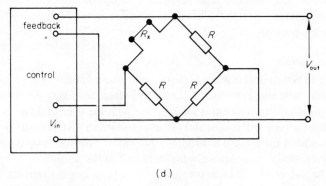

(d)

Figure 8.23. Hot wire anemometer and measuring circuits. (a) Anemometer
probe; (b) simple measuring circuit; (c) alternative form of constant
current operation; (d) constant temperature operation

for changes in the resistance of the hot wire (caused by the fluid flow), by varying the input level to the bridge and thus restoring the resistance of the hot wire to its initial value. In doing this the output voltage will be varied in a manner proportional to the air flow.

Humidity gauges

Humidity is a measure of the amount of water vapour present in a gas. It may be described in a number of different ways, but the most widely used expression is in terms of *relative humidity* which is the ratio of water vapour pressure present in a gas to water vapour pressure required for saturation of that gas at the temperature of measurement. The ratio is usually expressed as a percentage (% r.h.), and is temperature dependent.

Resistive humidity gauges may be classified as belonging to one of two types. One type is those having a resistive sensing element, that is a variation in the ambient humidity causes a variation in the resistance of the element that is usually a mixture of a hygroscopic salt[5], for example, lithium chloride and carbon on an insulating substrate between metal electrodes as shown in figure 8.24.

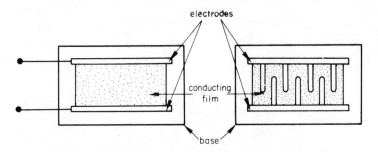

Figure 8.24. Resistive sensing element humidity transducers

The other form of resistance change humidity gauge is one in which the length change with humidity of a human hair or strip of animal gut is utilised to operate a displacement sensor which could be either potentiometric or strain gauge[12].

Photoresistive/conductive devices

The photoconductive cell has a conductivity which varies in accordance with the light intensity it receives. It consists either of a thin coating of a metal salt for example lead sulphide, or a single crystal of material such as doped germanium or cadmium sulphide, between conductive electrodes (to which connection wires are attached) and mounted on a glass plate. It is used in electrical circuits similar to those used for the hotwire anemometer (figure 8.23b and c) so that a signal dependent on light intensity is obtained. To form such a device into a transducer requires that the measurand shall modulate the light intensity falling on the photoconductive cell[6,13].

8.2 ELECTROMAGNETIC TRANSDUCERS

This type of transducer is generally selfgenerating, that is no external supply is required, its output voltage being produced by the movement of a magnetic flux field with relation to a coil system. The magnetic flux is usually derived from a permanent magnet, while the coil may be cylindrical and air cored or wound on to a core of ferrous material.

Linear velocity transducer

The simplest form of electromagnetic transducer is one in which a permanent magnet attached to an actuating shaft is free to move in a cylindrical coil (see figure 8.25), movement of the object/actuating shaft/magnet causing a voltage to be generated at the coil terminals the magnitude of which will be proportional to the velocity of the

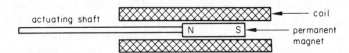

Figure 8.25. Electromagnetic linear velocity transducer

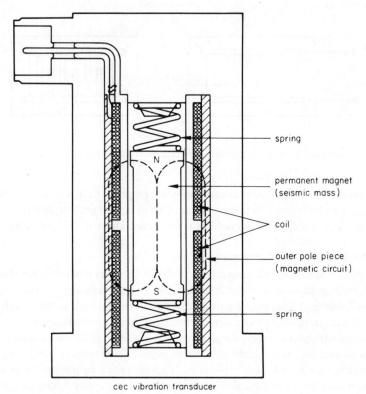

cec vibration transducer

Figure 8.26. Electromagnetic vibration transducer (Bell and Howell)

movement. A variation on this arrangement is shown in figure 8.26 where
the permanent magnet is supported between springs, and fitted with low friction
bearing rings. If such a transducer is attached to an object that is vibrating at a
frequency in excess of the low natural resonant frequency of the transducer, the
magnet would appear to an observer to be remaining stationary in space whilst
the case and coil oscillated around it, resulting in an output voltage, the magnitude
of which would be proportional to the amplitude of the vibration, and having a
frequency equal to that of the vibration. The relationships between sinusoidal
displacement, vibration, and acceleration are shown in figure 8.4.

Instead of the magnet being the moving member it is possible for the coil
to be moved, common examples being (a) some loudspeakers in which the coil
has a linear movement, and (b) a moving coil instrument where the motion is
rotary.

Angular devices

The use of transducers to determine angular motion has been widespread for
many years, for it includes devices such as the d.c. tachometer or tachogenerator,
in which the stator's magnetic field is derived either from a permanent magnet
or from a separately excited field coil, while the rotor winding (armature) is of
the form normally used in a d.c. generator fitted with a commutator. The output
voltage of a tachogenerator is proportional to rotary speed (or angular velocity)
it being about 5 V per 1000 rev/min for the permanent magnet tachometer, and
around three times this value for the separately excited transducer. In both forms
the polarity of the voltage is dependent on the direction of rotation.

If the rotor contains the permanent magnet field, then the magnetic interaction between it and the stator coils produces an a.c. voltage at the stator
terminals having a magnitude and frequency which are both proportional to
rotor speed. In many applications it is advantageous to operate with variations of
frequency rather than voltage level, since frequency is unaffected by circuit
impedance, loading, and temperature effects.

Toothed rotor tachometer

Tachometers using a toothed rotor made of a ferromagnetic material, and a
transduction coil wound around a permanent magnet are probably the commonest
form of frequency output angular velocity transducers. Figure 8.27 illustrates the
principle of operation of such a transducer, in which the magnet field surrounding
the coil is distorted by the passing of a tooth causing a pulse of output voltage
from the coil. If a multitoothed rotor is used, a pulse is generated for each tooth
that passes the magnet, and if six teeth are used the frequency of the output in Hz
will be 6 x rev/min/60 or rev/min/10. The r.m.s. value of the output voltage
will increase with a reduction in the clearance between the rotor and the pickup,
with an increase in the rotor speed, and with an increase in tooth size.

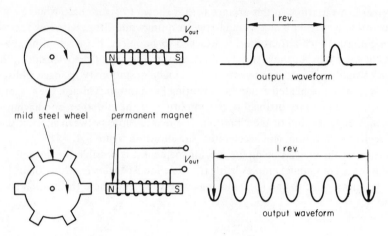

Figure 8.27. Operation of a toothed wheel tachometer

Magnetostriction transducers

When a ferromagnetic material is subjected to an alternating magnetic field, the domains within it vary in alignment; this causes a slight change in shape, and twice per cycle of magnetisation a piece of ferromagnetic material extends and contracts its length. This phenomena known as magnetostriction is reversible, that is if a ferromagnetic material is extended and compressed in a regular manner a voltage will appear at the terminals of a coil wound around it. This property is utilised in the construction of ultrasonic transducers[23], for such applications as wall thickness gauging, and underwater detection (sonar). Magnetostrictive transducers are rarely used outside these special applications.

8.3 REACTANCE CHANGE TRANSDUCERS

This is a group of transducers in which a displacement is used to modulate either a capacitive or inductive reactance. Since variations in reactance may only be measured when the supply is an alternating voltage (or current) the excitation source for reactance change transducers must be alternating. This may however make the transducer incompatible with the remainder of the measuring system, and to overcome this a number of reactance change transducers are manufactured with an internal d.c. to a.c. converter and an output demodulator so that both the transducer input and output are d.c. while the internal excitation is a.c.[4,6]

8.3.1 Capacitance Variation

It has been indicated that a change in almost any physical phenomenon may be converted to a displacement and as the capacitance between two parallel conducting plates is approximately proportional to $\epsilon A/d$, (where ϵ is the dielectric constant of the material between the plates of area A, separated by a distance d) varying any of these quantities will result in a change in the capacitance between the plates.

Variable dielectric

Figure 8.28 shows diagrammatically two types of dielectric variation transducer, one in which a dielectric sleeve is slid between coaxial electrodes to vary capacitance, and the other is a device for measuring the height of a liquid in a container. In both cases it is important that the dielectric constant of the material between the electrodes is substantially different from that of air so that an appreciable change in capacitance is obtained.

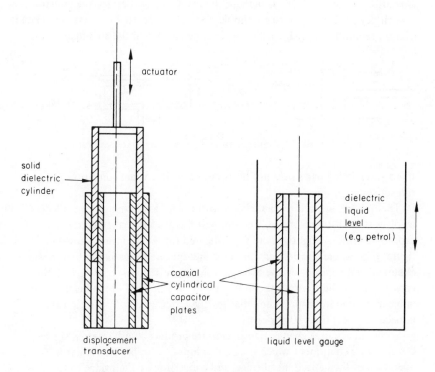

Figure 8.28. Variable capacitance transducers

Variable plate area

Transduction by variation of plate area may be performed by arrangements similar to those shown in figure 8.29, where linear or angular movement of the actuator causes one plate, or a set of plates, to vary the capacitance of the transducer.

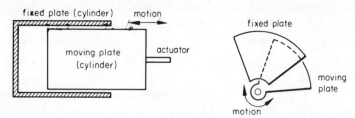

Figure 8.29. Capacitance transducers using plate area variation

Variable plate separation

The third form of variation, namely that involving a change in the electrode separation (see figure 8.30) is perhaps the most widely used for the construction of such devices is comparatively simple, the forces required to move one plate in relation to another may be made exceedingly small, and the alignment of the

Figure 8.30. Capacitance transducers using plate separation

fixed and moving parts may not be as critical as it is when some other forms of transducer are used.

The measuring circuitry associated with capacitance change transducers usually involves some form of bridge. The simplest form would have two resistive and two capacitive arms. A slightly more sophisticated form of bridge is shown in figure 3.25, where a ratio transformer is incorporated together with a resistance circuit to compensate for any leakage resistance effects associated with the capacitance of the transducer. Either circuit may be used in the balanced or unbalanced condition providing that for the latter mode of operation the unbalance is less than 10 per cent.

Some capacitance change transducers are produced with internal signal conditioning equipment which used a d.c. input to supply an oscillator, for operation of the capacitance bridge, and conversion of the bridge output to d.c.[4] Alternatively an oscillator circuit may be used so that a change in capacitance causes a change in output frequency.

8.3.2 Inductance Change Transducers

The inductance of a coil depends on the manner in which magnetic flux links its turns, thus by using suitable 'measurand to displacement' converters to cause a change in the magnetic flux linkages in a coil, a series of transducers exhibiting a change in inductance proportional to the measurand may be obtained. This change in inductance can be measured either as an amplitude change in a bridge balance, or as a change of resonant frequency in an oscillatory circuit. The methods by which this variation of inductance (flux linkages) is obtained are:

(a)   By changing the reluctance (or magnetic resistance) of the flux path by the movement of an armature, so that the displacement may be either linear or angular (see figure 8.31) this being incorporated in transducers to measure such quantities as pressure, acceleration, force, and displacement or position.

(b)   By the displacement of a slug of material having a high magnetic permeance that is constrained to move inside a centre tapped coil which is wound on a ferromagnetic core (see figure 8.32). When the slug is positioned centrally the inductance of the two halves will be equal, on movement to one side the inductance of one half will increase while that of the other decreases.

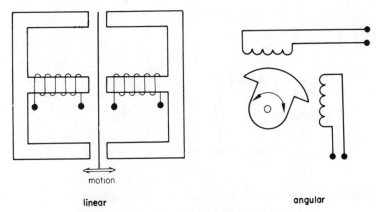

Figure 8.31. Variable reluctance transducers

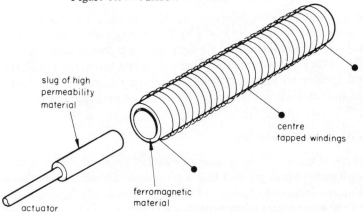

Figure 8.32. Variable permeance transducer

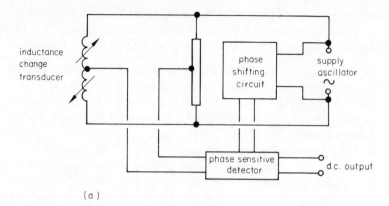

(a)

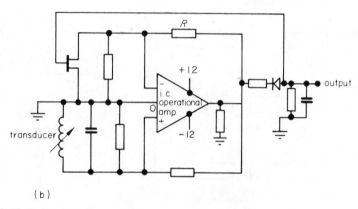

(b)

Figure 8.33. Inductive change transducer measuring circuits. Arrangement for
(a) d.c. output; (b) frequency output

The inductor may form the arms of a bridge circuit, or be incorporated in an
oscillator circuit (see figure 8.33a and b). The phase sensitive detector (see
page 32) is included so that the d.c. output has a sign corresponding to the
direction of the displacement. The phase shift circuit between the oscillator and
the p.s.d. is necessary because at zero displacement there will be a phase angle
between the reference voltage and the centre tap on the inductor (see also page
178).

8.3.3 Differential Transformer

In this type of transducer, the change of inductance is a variation of mutual
coupling between windings rather than a variation of self inductance. This variation of mutual inductance is obtained by the movement of a ferromagnetic core
within a coil arrangement of the type shown in figure 8.34. There are usually one
primary and two secondary coils. For a linear device the coils are wound on a

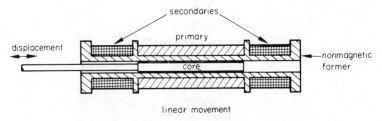

Figure 8.34. Differential transformer transducers

nonmagnetic and insulating former while in angular devices of this type a
ferromagnetic core for the coils may be used. The operational circuits used for
the differential transformer range from those such as that in figure 8.35a, which
has an a.c. output voltage that increases in magnitude for displacements either
side of the centre or zero position. This circuit can easily be modified so that a
d.c. output is obtained having a polarity dependent on the direction of displacement (see figure 8.35b). More complex arrangements are available commercially,
in which a d.c. input is converted to an a.c. voltage, that is varied by the transducer operation and then restored to d.c. so that the output voltage has a
magnitude and direction proportional to displacement.

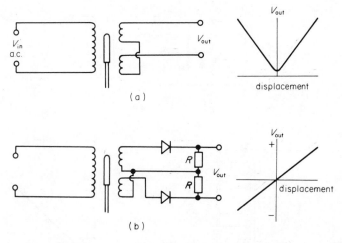

Figure 8.35. Operational circuits for use with differential transformer transducers

The differential transformer transducer is a durable, low impedance device,
having little or no friction, and capable of measuring total, steady state, and
dynamic displacements from 2 mm to 50 cm. It may therefore be used for
measuring a wide range of quantities if a suitable 'measurand to displacement'
converter is positioned between the quantity and the differential transformer.

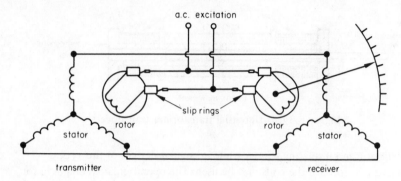

Figure 8.36. 'Synchro' measuring system

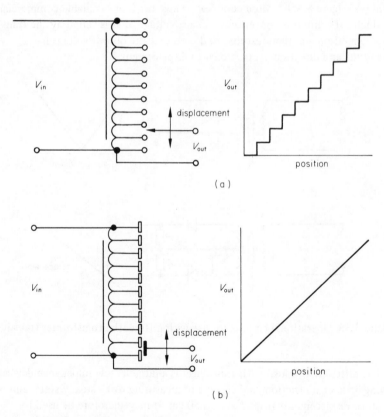

Figure 8.37. Types of autotransformer position transducer (a) direct contact;
(b) capacitively coupled

'Synchro'

A type of transducer used for the measurement and display of angular position or movement is the 'Synchro', in which two similar units are used, one termed the transmitter and the other the receiver[22]. Each is constructed with a two pole rotor and a stator having three windings distributed $120°$ apart (see figure 8.36). In operation the rotors of both units are fed, via slip rings, with an alternating current (usually 50 or 400 Hz) and if the rotors of the two units are in the same relative position to the stator windings no currents will flow between them. However, should the rotor of the transmitter be moved relative to its stator winding, the voltages induced in the two stators will be different and currents will flow in the connecting leads. These currents will produce a torque on the rotor of the receiver and cause it to move in sympathy with the rotor of the transmitter. The positioning accuracy of such a system is limited by the friction of the bearings in the receiver, and the calibration of the dial face attached to it.

Autotransformer transducer

A type of transducer that is difficult to classify is the autotransformer transducer, since it utilises electromagnetic coupling to divide an applied a.c. voltage[14]. To obtain a smooth variation in output voltage, capacitive coupling may be used (see figure 8.37b). This type of position transducer has a sufficiently good degree of resolution for it to be used in some machine tool control applications.

8.4 SEMICONDUCTOR DEVICES

The electrical properties of semiconducting materials are affected by variations in temperature, illumination, and force. The sensitivity of these semiconducting materials is very much greater than that of other materials, but this may introduce its own problems: for a semiconductor which, for example, is used in the measurement of strain may also be very temperature sensitive, and considerable additional circuitry will be required to compensate for such effects.

Thermistors

A very sensitive temperature transducer termed a thermistor may be manufactured by sintering oxides of such materials as manganese, nickel, cobalt, copper, iron, or uranium into tiny beads (0.1–2.5 mm dia.). Then for protection, coated in glass or mounted in an evacuated or gas filled bulb. Thermistors with resistance values of 10 Ω to over 100 MΩ may be obtained, and a typical thermistor will exhibit a *decrease* in resistivity of the order of 50 to 1 over a temperature range of 0 to $100°$C, and for the smallest versions can have a time constant of the order of 1 ms. The resistance variation is, however, exponential as opposed to the almost linear characteristics of a metallic resistance element.

A simple circuit would consist of a thermistor, a battery, and a microammeter calibrated in terms of temperature (see figure 8.38a) while a more sensitive arrangement would consist of a Wheatstone bridge arrangement (see figure 8.38b). This latter type of circuit is capable of determining temperature differentials as small as 0.001 °C. For some applications the thermistor is superior to other temperature transducers, making possible measurements that were not practicable prior to their discovery. They are a low cost device.

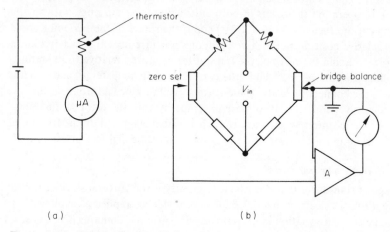

Figure 8.38. Circuits for measurement of temperature using thermistors

Semiconductor strain gauges

The materials used in the production of these devices are often termed piezo-resistive, meaning that when the gauge is subjected to a force there is alteration in the crystalline structure and a subsequent change in the electrical resistance. This variation in resistance is usually very much greater in a semiconductor than that obtained in a wire or foil strain gauge, and results in gauge factors of between 50 and 250 (+ or −) as opposed to the value of 2−3 for the resistance gauge. Typical semiconductor gauge arrangements are shown in figure 8.39, the gauge

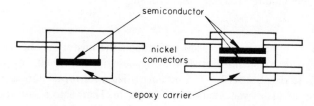

Figure 8.39. Semiconductor strain gauges

material normally being silicon, while the leads may be made from gold, silver, or nickel ribbon. Effective gauge lengths are typically between 2 and 7 mm.

Whilst the semiconductor strain gauge at first appears very attractive it does have severe drawbacks; first it is a nonlinear device, this effect may be reduced either by using heavily doped semiconductors, prestressing the strain gauge so that it operates over a linear portion of its characteristic, or by using compensation in the measurement circuit[6]. The second major drawback is the effect of temperature variations on the semiconductor, for the temperature coefficient of resistance materials used for these strain gauges is in the region of 60–100 times that of the materials used for resistance strain gauges. Therefore when semiconductor strain gauges are used temperature compensation must be incorporated in the measuring circuit. This may take the form of a 'dummy' gauge (see page 241) or be obtained by connecting a thermistor with the appropriate temperature/resistance characteristics in series or parallel with the semiconductor gauge, in each active arm of the bridge circuit (see figure 8.40) or by using a pair of gauges so that one is subjected to a positive strain and the other a negative strain.

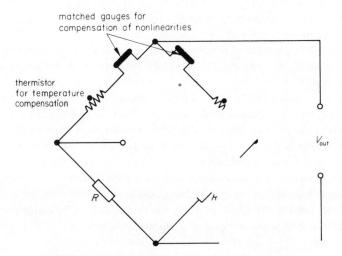

Figure 8.40. Semiconductor strain gauge bridge

Photodiodes and phototransistors

The electrical characteristics of most diodes and transistors are affected by light, hence it is normal to prevent light from acting on the semiconductor junction of these devices, by enclosing it in a screening can or other opaque container. If this container is removed from a transistor (or diode) incident light falling on the junction will affect the output of the device. Photodiodes and phototransistors are constructed so that this normally undesirable property is enhanced. The majority of light sensing semiconductors are silicon, but when optimum response in the infrared region is required germanium devices may be used. Both diodes and transistors are usually sealed in a metal can, there being a lens or window fitted to admit light to the junctions. Time constants of the order of 1 μs or less are common.

8.5 PIEZOELECTRIC TRANSDUCERS

These devices are those which utilise the piezoelectric characteristics of certain crystalline and ceramic materials to generate an electrical signal. The piezoelectric effect was discovered by Pierre and Jaques Curie in 1880 when they found that by placing weights on to a quartz crystal an electrical charge could be generated. Subsequent researchers have shown that there are about forty crystalline materials that when subjected to a 'squeeze' (the Greek work 'piezein' means to squeeze) an electrical charge is generated. The Curie brothers also discovered the reverse effect, that is a dimension change is obtained when an electric stress is applied to a quartz crystal.

The crystal materials are of two basic types, natural crystals and synthetic crystals, the latter type being mainly ceramic 'crystals' of which barium titanate was the first to be commercially employed. The addition of controlled impurities such as calcium titanate was found to improve some of the 'crystal's' characteristics. Research on this and other synthetic 'crystals' has led to them being used more frequently than natural crystals in the production of piezoelectric transducers.

The force applied to the crystal may be that of shear, compression, or bending, depending on the particular application for which the transducer has been designed. The merits of operating the crystal with the various force modes will not be discussed here as they are described in detail elsewhere[1,2]. It must, however, be remembered that since piezoelectric transducers belong to the selfgenerating category their use is predominantly in dynamic applications, a common form being as seismic accelerometers; although the piezoelectric principle may be effectively used in any situation where the measurand can be converted to a force that results in the stressing of a suitable crystal.

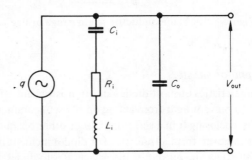

Figure 8.41. Equivalent circuit of a piezoelectric transducer

All piezoelectric crystals have a high output impedance, and for this reason the instrumentation to which the piezoelectric transducer is connected must have a high input impedance, or for most applications a charge amplifier (see page 175), must be connected between the transducer and the display instrument. A disadvantage is the additional expense involved. The equivalent circuit of a piezoelectric transducer is shown in figure 8.41, and this in practice may for simplicity be considered as a voltage source whose series impedance is a capacitor having a value of a few picofarads. The internal resistance (R_i) will normally exceed 20 GΩ and can thus be ignored when considering the overall performance of the transducer, similarly the effects due to the lateral inductance are beyond the upper frequency limit of the transducer and can also be ignored.

8.6 DIGITAL TRANSDUCERS

True digital transducers are those whose output is represented by a number of discrete increments. The term is generally used, however, to include those devices which have a pulse output that can be applied to a digital counter. Examples of this latter type are the toothed rotor tachometer, and a similar effect obtained by using photoelectric transduction, for example figure 8.42 where light shines

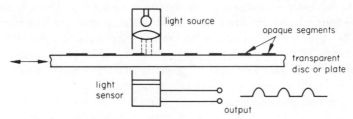

Figure 8.42. Photoelectric transducer for measurement of motion

through a segmented disc or plate which, when in motion, modulates the light falling on to the light sensor. The devices that generate a single pulse train cannot, in general, give an indication of direction; but this difficulty may be overcome for the toothed rotor tachometer by using two sensing heads positioned relative to each other so that their electrical outputs have a phase displacement that will be lagging or leading depending on the direction of movement of the rotor. In the case of the photoelectric transducer determination of direction may be performed by the addition of a second track of opaque segments with its own illumination sensor. A higher degree resolution in photoelectric transducers may be obtained by using interference pattern techniques derived from the use of optical gratings[14,15]. A number of different pattern types exist, but the most widely known is the 'Moiré' pattern. To obtain this a strip with alternate parallel opaque and transparent segments is moved past a similarly striped plate that is set at an angle to the moving plate. Positioning a line of light sensors on one side of the stripes and shining light through the stripes on to them results in a variation in the particular sensors illuminated, or obtains considerable variations in the sensor's output level with the movement of the striped strip. The outputs from such arrangements will approximate to a sine wave, the frequency of which will be dependent on the number of segments on the strips.

Quartz crystal thermometer

Quartz crystals are sensitive to temperature and this property may be utilised to measure temperatures in the range 230–500 K by synthetic quartz crystals specially designed to have a linear temperature coefficient of frequency. Connecting such a crystal into an oscillator circuit, it can exhibit a frequency–temperature slope[16] of as much as 1 kHz/°C, and if the oscillator output is mixed with a reference oscillator a beat frequency may be obtained that may be displayed on a frequency counter as a temperature.

Investigations into the use of crystalline oscillations for the determination of humidity are in progress, and promising results have been obtained by using a hydroscopic coating on a quartz crystal.

Digital encoder

The above transducers are all pulse producing devices, capable of detecting a change in the measurand rather than describing its absolute or unique state. In addition to this latter requirement a true digital transducer should have an output that is in a form suitable for direct entry into a digital computer or data handling system. This latter type of transducer is known as a digital encoder and may read continuously or be momentarily stopped to obtain a readout (depending on the design and code used). Encoders are generally an integral part of a more complex transducer, for example a wind direction indicator.

While digital encoders may be made for linear applications one of the commonest is the shaft position encoder. It consists of a disc or drum with a digitally coded scale which may be formed either from a combination of translucent and opaque segments, or a combination of conducting and insulating surfaces, that are coded so that there exists a unique form for each discrete position on the disc as shown in figure 8.43. Such discs are manufactured with diameters from 5 cm to 25 cm

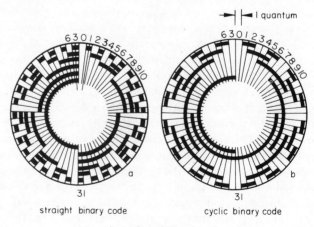

comparison of straight and cyclic codes

Figure 8.43. Shaft encoders (Courtesy of Vactric Control Equipment Ltd)

and give unique codes for between 100 and 50 000 positions per 360°. These coded positions are read either by a series of brushes or by a light source and sensor arrangement, and fed into an appropriate number of channels. The sequence and order of indications represent the shaft in the coded form. A special code known as the Gray code is commonly used in the operation of these devices.

8.7 THERMOELECTRIC TRANSDUCERS

In 1822 John Seebeck reported that if a magnetic needle was held near a circuit made from conductors of two different materials, the magnetic needle was deflected when heat was applied to part of the circuit, thus implying the presence of an electric current[18]. In 1834, a French watchmaker, Jean Charles Peltier, found that when a current was passed through a junction of two different conductors it changed temperature. In 1854 William Thompson (Lord Kelvin) discovered that if a current carrying conductor is heated at one point along its length, points at equal distances on either side of the heat source along the length of the conductor will be maintained at different temperatures, points on the negative potential side of the heat source being lower in temperature than those on the positive potential side.

Thermocouple

The Seebeck effect, which is a combination of the Peltier and Thompson effects, is utilised in what is probably the most widely used method of thermometry, namely the thermocouple.

These transducers consist of a pair of bars or wires of dissimilar metal joined at both ends: one end is used as the hot (sensing) junction, while the other end is used as a cold or reference junction, as shown in figure 8.44. The term

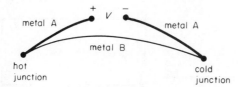

Figure 8.44. Basic thermocouple circuit (see also figure 1.16)

thermocouple has through usage come to mean a single thermojunction, consisting of two lengths of dissimilar wire, insulated from each other but joined at one end. In conforming with this usage it becomes necessary to describe the combination of a 'hot' and 'cold' junction as a thermocouple circuit.

A suitable cold junction may be obtained by the use of melting ice, but such a reference is not always convenient or for that matter necessary, for example if a thermocouple is being used to measure a temperature that is of the order of 1000 K, variations of 5°C or so in the temperature of the reference junction will have only a small effect on the accuracy of the measurement. Thus ambient air temperature may satisfactorily be used as a reference for some applications.

Output measurement

A thermocouple may be used as a voltage source, when its output should be measured by a high impedance instrument, for example a digital voltmeter or a d.c. potentiometer, the displayed millivolts being dependent on the unknown temperature. Alternatively a low impedance instrument may be used to complete the circuit as in figure 8.45 when the magnitude of the current in the circuit

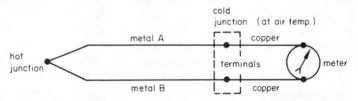

Figure 8.45. Commonly used thermocouple circuit

will be dependent on the magnitude of the unknown temperature. When temperatures near that of ambient are to be measured with a thermocouple and it is inconvenient to use a fixed temperature reference junction, compensating circuits must be incorporated in the measuring system. One method of arranging automatic compensation is shown in figure 8.46 where a temperature sensitive bridge is included in the thermocouple circuit, such that variations in the ambient temperature level are compensated for by the changes in the resistance

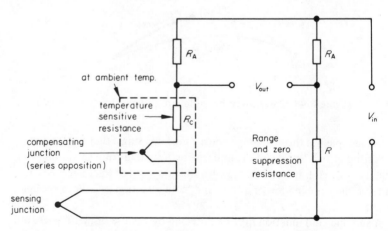

Figure 8.46. Thermocouple with bridge type compensation

R_c and the compensating junction. Multichannel thermocouple systems sometimes use a heated, thermostatically controlled reference junction (see figure 8.47)
where by the use of regulated temperature control of the reference junction, a
system accuracy of ± 0.3°C may be maintained. When the reference junction is
not held at 0°C, the observed value must be corrected by adding to it a voltage

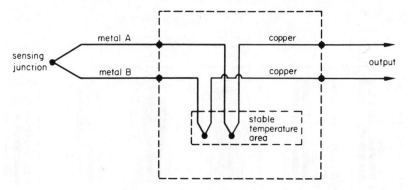

Figure 8.47. Thermocouple reference junction

that would have resulted from a temperature difference equal to the amount by
which the reference junction is above 0°C, that is $E_T = (E_t + E_r)$ where E_T is
the total e.m.f. at temperature T, E_t is the e.m.f. due to the temperature difference
between the sensing and reference junctions, and E_r is the e.m.f. due to the temperature of the reference junction above 0°C. Since the temperature—voltage characteristic of a thermocouple is nonlinear, it is important that temperatures are determined
by the above process (rather than by converting an e.m.f. to a temperature and then
adding this to the ambient temperature), also the reference junction should be
maintained at, or near, the temperature used during calibration of the thermocouple or errors will result.

Thermocouple materials

The materials most commonly used to form thermocouples are:

copper—constantan, 0—370°C (270—640 K)
iron—constantan, 0—760°C (270—1030 K)
nickel/chromium—nickel/aluminium, 0—1260°C (270—1500 K)
tungsten—rhenium, up to 2760°C (3000 K) and
platinum—rhodium alloy which may be used up to 1750 K, but is also used to
define the International Temperature scale from 800—1340 K (see also reference
21).

The conductors of a thermocouple must be insulated from each other, and
except at low temperatures, it has become the practice to use mineral insulation
and sheath the conductors and insulator with stainless steel. It is possible to

purchase thermocouples in a variety of sizes, from 0.25 mm to 3.0 mm dia., and shape of sensing junction (see figure 8.48). In addition to this the locating of the measuring junction is important, considerable care being necessary to ensure that the temperature recorded is the unaffected temperature of the measuring point[17].

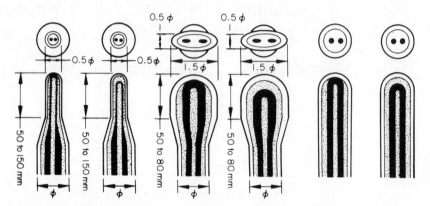

Figure 8.48. Hot junction shapes — mineral insulated metallic sheathed thermo-couples (Pye—Unicam Ltd)

When several thermocouples made from the same materials are connected in series the hot junctions all being at one temperature and the cold junctions at another, they are said to form a thermopile. The output voltage from a thermopile is equal to the output voltage from a single thermocouple multiplied by the number of thermocouples in the thermopile assembly, providing that the materials are connected in the correct sequence and the reference junctions are all kept at

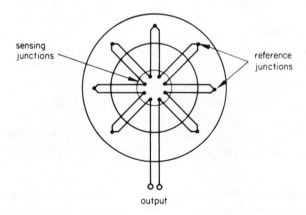

Figure 8.49. Thermopile formed into a radiation pyrometer

the same temperature. If the thermopile is arranged in the configuration shown in figure 8.49 so that the hot junctions are at the focal point of an optical system, then the device so produced forms the sensor for a thermal radiation pyrometer, and may be used to determine the temperatures of heated surfaces without physical contact[11].

The advantages of nonconducting methods of temperature measurement have led to the development of infrared thermometers that use vacuum thermocouples, metal film bolometers, or photodiodes as the detector of radiant energy. Such instruments vary in complexity from portable, hand held thermometers, which

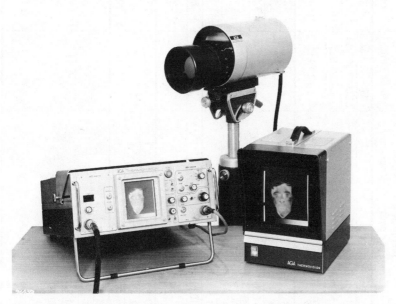

Figure 8.50. An infrared thermal imaging equipment showing the temperature profiling on a face (Courtesy AGA (U.K.) Ltd)

may be used to determine the temperature at a point on the surface of an object, to the sophisticated thermal imaging systems in which the focal point of measurement is scanned across the surface of the object under investigation. The output from the pyrometer head in this type of instrument is used to modulate the brightness of a spot that is scanning the screen of a cathode ray tube enabling a 'television type' picture of temperature variations to be built up [19,20,24]. This type of equipment can be used to detect temperature differences as small as 0.1°C and has many applications both in the industrial and medical fields. Figure 8.50 shows such an equipment displaying the temperature pattern obtained from a face.

Table 8.2. Comparison table of some common transducers

type	measurand	range	resolution (per cent)	typical errors (per cent)	impedance	output	frequency range
potentiometric	acceleration	$1 \to 100$ g	0.5	± 1	17–23 kΩ	2 mA max	d.c. → 0.1 Hz
	displacement	$0 \to 1$ cm $\to 30$ cm	0.05–0.1 mm	± 0.1	500–200 kΩ	10 mA max	d.c. → 0.1 Hz
resistance	force (load) / pressure	10 N $\to 1$ kN	0.2	± 0.5	1–20 kΩ	10 mA max	
	pressure	$0 \to 300$ kN/m² $\to 7$ MN/m²	0.3	± 0.5	1–20 kΩ	10 mA max	
resistance strain gauge / transducer	acceleration	$5 \to 200$ g	limited by measuring circuit	± 1	120 Ω	± 3.5 mV/V	1 Hz → 10 kHz
	displacement	$0 \to 5$ mm				± 3.5 mV/V	
	force (load)	± 0.5 N $\to 1.6$ MN			–	± 3.5 mV/V	
	pressure	$0 \to 600$ kN/m² $\to 14$ MN/m²			1 kΩ	± 1.5 mV/V	
	temperature	$270 \to 1000$ K				± 1.5 mV/V	
variable reluctance	displacement	$0 \to 1$ mm $\to 20$ cm	ditto	± 1	100 Ω	4–20 mV/V	1 Hz → 1 kHz
	pressure	$0 \to 30$ kN/m² $\to 35$ MN/m²		± 0.5	1 kΩ	4–20 mV/V	
piezoelectric	acceleration	$2 \to 20\,000$ g	ditto	± 1	200 pF	35 mV/g	1 Hz → 1 MHz
	force (load)	$0 \to 2$ MN		± 1	–	4 pC/N	
	pressure	$0 \to 40$ MN/m²		± 1	2.5 nF	0.3 pC/kN/m²	
resistive semi-conductor (transducer)	acceleration	$5 \to 200$ g	ditto	$\pm 1 \to 2$	120 Ω	2–10 mA	d.c. → 1 kHz
	displacement	$0 \to 5$ mm					
	force (load)	$0.5 \to 1.6$ MN					
	pressure	$0 \to 600$ kN/m² $\to 14$ MN/m²			1 kΩ		
	temperature	$250 \to 500$ K and small temp. diffs.					
thermocouple	temperature	$50 \to 2500$ K	ditto	$\pm 1 \to 3$	10–30 Ω	40 V/k	d.c. → 0.1 Hz

Frequency range axis: d.c., 0.1 Hz, 1, 10, 100, 1 kHz, 10, 100, 1 MHz, 10

8.8 COMPARISONS

In table 8.2 are listed the main characteristics and applications of some of the more common transducers. The values quoted should only be taken as a general guide, for special purpose transducers of each type may be manufactured with characteristics very different to those quoted in the table.

REFERENCES

1   C. M. Crede and C. E. Harris. *Shock and Vibration Handbook,* McGraw-Hill, New York (1961)

2   Dale Pennington. *Piezoelectric Accelerometer Manual.* Endevo Corporation, Pasadena (1965)

3   T. Potma. *Strain Gauge Theory and Application,* Iliffe, London (1967)

4   R. H. Cerni and L. E. Foster. *Instrumentation and Engineering Measurements.* Wiley, New York (1962)

5   K. Arthur. *Transducer Measurements.* Tektronix Inc., Beaverton (1970)

6   H. N. Norton. *Handbook of Transducers for Electronic Measuring Systems,* Prentice Hall, Englewood Cliffs, N.J. (1970)

7   J. L. Thomas. 'Precision Resistors and their Measurement', pp. 152–183, *Precision Measurements and Calibration.* Ed. F. L. Hermad and R. F. Dzruba. *N.B.S. Special Publication 300 Vol. 3.*

8   J. J. Hill and A. P. Miller. 'An Inductively Coupled Ratio Bridge for Precision Measurements of Platinum Resistance Thermometers.' *Proc. I.E.E.* **110**, No. 2 (1963)

9   F. E. Smith. 'On Bridge Methods for Resistance Measurements of High Precision in Platinum Thermometery.' *Phil. Mag.* (6) **24**, 541 (1912)

10   F. G. Brickwedd (Ed. Part 1) and A. I. Dahl (Ed. Part 2). *Temperature, its Measurement and Control in Science and Industry.* American Institute of Physics, Reinhold, New York (1962)

11   H. D. Baker, E. A. Ryder, and N. H. Baker. *Temperature Measurement in Engineering, Vol II.* Wiley, New York (1961)

12   C. Watchorn. 'Relative Humidity.' *Cambridge Technical Review* (August 1968)

13   'Optical Techniques of measurement in Control.' *I.E.E. Colloquium Digest.* No. 1970/13.

14   R. C. Brewer. 'Transducers for Positional Measuring Systems. A review of progress.' *Proc. I.E.E.* **110**, No. 10 (Oct. 1963)

15   D. K. Ewing. 'Measurement of Speed Transients using Optical Gratings.' *N.E.L. Report No. 247* (1966)

16   H. C. Stansch. 'A Linear Temperature Transducer with Digital Output.' *19th Annual Conference. New York* (Oct. 1964)

17   H. D. Baker, E. A. Ryder and N. H. Baker. *Temperature Measurement in Engineering, Vol I,* Wiley, New York, (1953)

18   'Thermoelectricity.' *Colliers Encyclopedia, Vol 22* (1965)
19   D. B. Lloyd. 'An Introduction to Thermal Imaging.' *Electronic Equipment News* (July 1968)
20   J. Norrin. 'Thermal Imaging.' *Electronic Equipment News* (Sept. 1970)
21   *Thermocouple Reference Tables*, B.S. 4937:1973.
22   A. R. Upson and J. H. Batchelor. *Synchro Engineering Handbook.* Hutchinson, London (1966)
23   M. L. Gayford. *Electroaccoustics,* Butterworth, London (1970)
24   D. F. Cassidy. 'Age of Temperature Monitoring by Non-contacting Methods.' *Control and Instrumentation,* pp. 26–27 (Feb. 1972)

9

Data Recording and Analysis

Consider a researcher investigating a piece of equipment such that readings of temperature, pressure, movement, voltage, current, etc., are required to be recorded at regular intervals over a period of time to establish consistency of operation. Until the advent of the data recorder this type of situation required attendants to take the measurements. This used to be done somewhat laboriously and at times with dubious accuracy. However, now, when some form of data recorder is used, readings may be made with known limits of error either as a continuous record or in a predetermined sequence at controlled time intervals over an extended period.

A form of data recorder, described earlier in this book, is the multichannel pen recorder (see page 46). This is a very useful instrument for recording slowly varying quantities and presenting the result in graphical form.

9.1 DIGITAL RECORDING SYSTEMS

The merits of using digital methods have been listed in chapter 6. One feature of some of these instruments is an output that can be used to operate a printer or paper punch. This feature enables a data recording system to be built around the digital instrument. The basis of accomplishing this is shown in figure 9.1 where a number of analog signals are conveyed via a scanning switch (multiplexer) to an analog to digital converter (d. vm.) the output of which is fed via a drive unit to

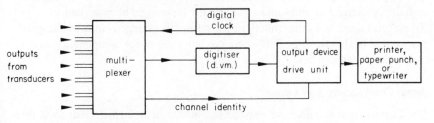

Figure 9.1. Block diagram of the basic data logger

some form of printer or paper punch[1,2]. The digital clock provides, at a pre-selected time interval (for example 10 min, 1 h, etc.), a command pulse to initiate the scan of the analog inputs. Simultaneously, the output device would be supplied with a signal from the clock so that the time at the start of the scan was recorded. The magnitude of the signal on each channel is identified in the record with a particular channel by recording the channel number (channel identification) adjacent to the signal magnitude.

The arrangement shown in figure 9.1 indicates the important components of a data logger built around a digital voltmeter, and whether the logger under consideration is a permanent installation or a small unit used to assist laboratory investigation, the logger output will be of greater value if it is presented in units appropriate to the measurand rather than as voltages which have to be operated on to obtain a correct interpretation. This desirable condition may be obtained by means of a few additions to the basic logger. For example, compensation for ambient temperature variations may be made to the output from thermocouples, and the input to the digitiser modified so that the print out is in °C.

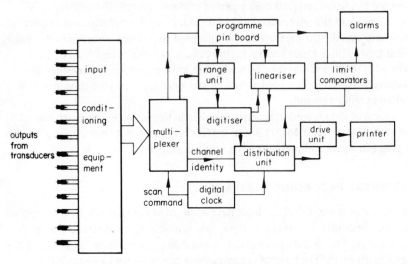

Figure 9.2. A comprehensive data logging system

Similarly, the outputs from other transducers can be scaled so that the logger output is presented in units appropriate to the measurand of each channel. To perform these and other functions the schematic arrangement of the logger is modified to that in figure 9.2, where the functions of the component parts shown are as detailed below.

Input Conditioning Equipment

The output from a transducer may be a.c. or d.c.; if it is the former some form of a.c. to d.c. converter must be inserted between the transducer and the logger, because the input to a logger that uses a d. vm. as a digitiser should normally be in the range 0–50 mV d.c., although the upper and lower limits of voltage which are normally imposed by the scanner are of the order of 100 V and 10 μV respectively. Voltages outside these limits must be scaled, or amplified according to their magnitudes. Currents must be converted to voltage drops, and pulse or variable frequency signals operated on by suitable circuits to produce voltage inputs compatible with the logger input requirements. The treatment of the inputs at this stage is purely on an individual basis and may be expensive.

Input Scanning Switch or Multiplexer

The idea of multiplexing originated on the telegraph systems where one line was time shared so that several morse coded messages could be sent 'simultaneously' by connecting appropriate pairs of input and output terminals to the line for 5 s periods[3].

In a data logger, the multiplexer is a time controlled switch that selects, in a predetermined order, the signals to be logged. It may use diode switching or reed relays, the former type being characterised by a very high speed capability together with high cost, thus restricting its use to those applications where speed is the overriding consideration. The electromechanical types have superior switching characteristics, and are capable of switching speeds up to 100 channels per second. The switching of both types is usually time controlled by the digital clock.

Program Pinboard

This is used to route the signals on each channel through appropriate paths. It is a feature that is not included in all data loggers but may be very useful in reducing the restrictions on the input signals and thus reducing the amount of scaling required to be performed on each channel at the input conditioning stage.

Lineariser

Reference was made earlier to the scaling of inputs so that the printout is in units appropriate to the measurand. Many transducers have an output voltage that is linearly related to the measurand, and for these devices a voltage divider of appropriate ratio will perform the necessary function. However, the output characteristics of some devices, for example a thermocouple, are nonlinear and routing the output from such a device via an appropriately programmed circuit facilitates the linearisation of the signal together with the conversion from mV to °C.

Digital Clock

The digital clock (see page 196), synchronised to the power supply or to an internal crystal oscillator for greater accuracy), governs the timing throughout the system by applying triggering pulses to the various parts of the system, thus initiating the sequences of events to record each signal. The clock is also used to supply the signals whereby the time of day, or lapsed time, may be recorded.

Limit Detectors

To facilitate a certain amount of protection in the automatic system a series of limit detectors may be incorporated. These units compare the signal magnitude of selected channels with a preset voltage; should the signal be outside the given limits it can be recorded (for example by red print) and also activate alarms or protective devices (that is, switch off).

Output Devices

The permanent record may be obtained from a paper tape punch, a strip or line printer, an electric typewriter, or on magnetic tape. The selection of the most suitable device will be governed by cost and the required speed of operation. The cheapest printer will have a maximum speed of operation of 2 or 3 channels per second, whereas with line or block printers much greater speeds are possible.

Other Features

Some data loggers incorporate facilities for supplying and using such sensors as strain gauges, and resistance thermometers; also reference junctions for thermocouples. Loggers are available in several forms, varying from small modular units which are very suitable for research/development investigations, to large purpose built units for installation in power stations, processing factories or on board ship. For example, a unit installed on board the merchant ship 'Manchester Challenge' scans 200 inputs every half second for alarm purposes and prints the 'Engineer's Log' once every four hours, printing one channel in each scan.

The development of low cost 'mini' computers has led to a new generation of data loggers in which higher scanning speeds are possible, and more sophisticated scaling or operating functions are available; the logger constitutes part of the 'on-line' control of the system whose performance is being recorded.

The above remarks have all been concerned with the data logger built around an A–D converter. An alternative scheme is one in which the inputs to the logger are signals whose frequency is dependent on the measurand[4,5]. Some measurands can conveniently be obtained as a frequency (see page 253), the most obvious example being rotary speed, while other measurands may require the use of voltage to frequency converters (that is, voltage controlled oscillators). With the expansion in use of integrated circuits it has become a practical proposition to incorporate a voltage to frequency converter within the transducer. The major advantages of a system based on the measurement of frequency are that the problems of signal attenuation during transmission from source to logger are eliminated, and that a reliable frequency to digital convertion is cheaper than a d. vm. of comparable accuracy. The disadvantages of the system are that it may be necessary to provide analog voltage to frequency equipment for each channel, and that 'crosstalk' between channels may introduce errors (see page 148).

Data loggers are sampling devices, that is, the magnitude of the signal on each channel is briefly 'looked at' and recorded on each scanning cycle, and care must be exercised to ensure that the scanning rate is sufficiently fast for all the variations in the measurand to be reconstructed from the record. The limiting requirement of sampling is that the frequency of sampling[6] shall be at least twice the highest frequency component of the signal being sampled in order that it may be reconstructed in its original form. An alternative method of looking at this is to consider a system which has a pass band of (say d.c. to f Hz, then the sampling frequency must be at least $2f$. However, the sampling theory assumes the presence of signal reconstruction equipment and a more practical sampling rate for a data logger is six times the highest frequency component of the signal[7].

The development of high speed A–D converters, together with the availability of low cost digital memory circuits, have led to the transient recorder. One instrument of this type, manufactured by Datalab, has a 5 MHz, 8 bit A–D converter connected to an 8 bit x 1024 word MOS shift register memory. This capability results in a recording cycle that stores 1000 equally spaced samples over a time interval that is adjustable from 200 μs to 10 s in a 1–2–5 sequence. The digitally stored information may be displayed on an oscilloscope or used as an input signal for either a pen recorder or a paper punch (via a suitable interface); this last facility enables computer analysis of transients to be made.

9.2 INSTRUMENTATION TAPE RECORDERS

The first magnetic recording of information was performed by Valdemar Poulsen with his 'Telegraphon' in 1893, in Denmark. By 1935 the Germans had successfully developed a suitable plastics 'tape' although in Britain and the U.S.A. steel wire or tape was used until around 1950 when the advantages of plastics tape over wire became widely accepted and wire recording became obsolete. Advances in materials technology have led to the tape recorders of today having a performance many time superior to that of the 'Telegraphon' but the magnetic principles remain the same.

Tape recorders consist of three basic parts:

(a) *A recording head.* A device that responds to an electrical signal in such a manner that a magnetic pattern is created in a magnetisable medium. Its construction (see figure 9.3), is similar to a transformer with a single winding, the signal current flowing in this winding and producing a magnetic flux in the core material. The magnetic coating on the tape bridges the nonmagnetic gap in the 'head' core and if the tape is moving past the gap the state of magnetisation of the oxide as it leaves the gap will be retained, thus the actual recording takes place at the trailing edge of the gap.

(b)   *The magnetisable medium.* 'Magnetic tape' is composed of a coating of fine magnetic oxide particles on a plastics ribbon. The oxide particles conform to and retain the magnetic pattern induced in them by the recording head. Associated with the tape will be the tape transport system for precise control of tape movement (see page 288).

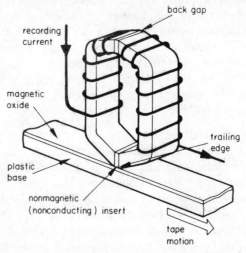

Figure 9.3. Simplified diagram of a record head

(c)   *The reproduce head.* This device detects the magnetic pattern stored on the tape and translates it back into the original electrical signal. The reproduce head, whilst similar in appearance to the record head, is fundamentally different in operation.

Consider the magnetised tape travelling across the gap of a head such as that shown in figure 9.3; to induce a voltage in the winding requires that there be a change in the level of magnetisation on the tape, that is $e_{rep} \propto N \, d\phi/dt$, N being the number of turns in the winding of the reproduce head. Since the voltage in the reproduce head is proportional to $d\phi/dt$, the reproduce head acts as a differentiator.

For example, if the signal to be recorded is $A \sin \omega t$, both the current in the record head and the flux in its core will be proportional to this voltage. Assuming that the tape retains this pattern and regenerates it in the reproduce head core, the voltage in the reproduce head winding will be

$$e_{rep} \propto \frac{d\phi}{dt}$$

and $\qquad \dfrac{d\phi}{dt} = \omega A \cos \omega t$

so $\qquad e_{\text{rep}} \propto \omega A \cos \omega t$

and the signal out of the reproduce head is the derivative of the input. In addition to this its magnitude is proportional to the frequency of the input. Thus to maintain amplitude fidelity these factors must be compensated for by the characteristics of the output electronics. This process is known as equalisation, the overall output characteristic being shown in figure 9.4.

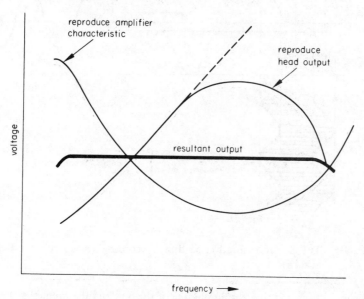

Figure 9.4. Characteristics of the components of the reproduce system to obtain amplitude fidelity (Adapted from Hewlett Packard)

In instrumentation recording, three techniques are in general use: direct, frequency modulated, and pulse duration modulation. Comparison of the three methods indicates that while the first method may have the greatest bandwidth, smaller errors are possible with the other methods, the frequency modulation recording process having the greatest number of applications in instrumentation.

9.2.1. Direct Recording

While direct recording can be used for instrumentation purposes it is more usually used for the recording of speech and music. It is the simplest method of recording and usually requires one tape track for each channel[8,9]. The signal to be recorded is amplified, mixed with a high frequency bias, and fed directly to the recording head as a varying electric current. The bias is introduced in order to eliminate the nonlinear effects of the tape material's magnetisation curve. Both the amplitude and frequency of the bias is selected to be several times larger than the maximum amplitude and highest frequency contained in the input signal. The result of this is that the intensity of the magnetisation is proportional to the instantaneous amplitude of the input signal (see figure 9.5).

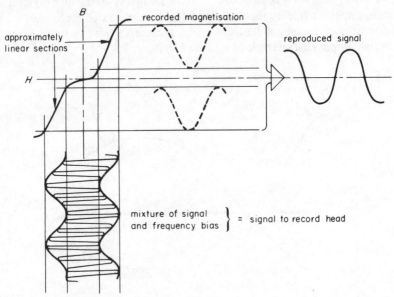

Figure 9.5. Classical representation of 'direct recording' (Adapted from Hewlett Packard)

It should be noted that the combining of the bias and the input signal is accomplished by a linear mixing process that is not an amplitude modulation process.

The advantage of the direct record process is that it provides the greatest bandwidth obtainable from the recorder. It also requires only simple, moderately priced electronics. However, since a signal is induced in the reproduce head only when there are variations in the level of magnetisation on the tape as it passes this head, the low frequency response of a direct record system does not extend to d.c., the limit being around 50 Hz. The upper frequency limit being around 2 MHz at a tape speed of 3.05 m/s (120 in/s). The direct recording process is also characterised by some amplitude instability caused by random surface inhomogenities in the tape coating. At long wavelengths (low frequencies) the amplitude variations caused by this may only be a few per cent; however for frequencies near the upper bandwidth limit for a given tape speed, amplitude variations can exceed 10 per cent and momentary decreases of over 50 per cent (called 'dropouts') may occur. This also indicates one of the difficulties associated with magnetically recording digital data.

Direct recording should therefore only be used when maximum bandwidth is required and amplitude variations are not unacceptable. If maximum bandwidth with relative freedom from dropouts is required, a single input signal can be recorded on several channels simultaneously.

In sound recording the ear will average amplitude variation errors, and whilst the audio tape recorder utilises the direct recording process, it is seldom satisfactory to use an audio recorder for instrumentation purposes. The former is designed to take advantage of the rather peculiar spectral energy characteristics of speech and music, whereas an instrumentation recorder must have a uniform response over its entire range.

9.2.2. Frequency Modulated Recording

Frequency modulation overcomes some of the basic limitations of direct recording at the cost of reducing the high frequency response. The bandwidth of the recording may be extended down to d.c. and the reproduced signal is not significantly degraded by amplitude variation effects. In the f.m. recording system, a carrier oscillator is frequency modulated by the level of the input signal. That is, a particular frequency is selected as the centre frequency and corresponds to zero input signal. Applying a positive voltage input will deviate the carrier frequency a specified percentage in one direction, whereas the application of a negative input voltage would deviate the frequency by an appropriate percentage in the opposite direction (see figure 9.6). Thus d.c. voltages are presented to the

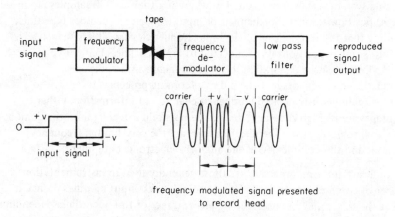

Figure 9.6. Schematic representation of 'frequency modulated recording'

tape as frequency values, and a.c. voltages as continually changing frequencies, and amplitude instabilities will have little or no effect on the recording.

On playback, the reproduce head output is demodulated and fed through a low pass filter which removes the carrier and other unwanted frequencies generated in the modulation process. In a typical f.m. recording system, ± 40 per cent deviation of the carrier frequency corresponds to plus and minus full scale of the input signal.

Whilst f.m. recording overcomes the d.c. and amplitude limitations of the direct recording process its own limitations are:

(a)   limited (20—50 kHz) high frequency response compared to the direct record process
(b)   dependence on the instantaneous tape velocity
(c)   associated electronics that are more complex and therefore more costly than those required for the direct process.

9.2.3. Pulse Duration Modulation Recording

Recording by pulse duration modulation (p.d.m.) is a process in which the input signal at the instant of sampling is converted to a pulse the duration of which is made proportional to the amplitude of the input signal at that instant. For example, in recording a sine wave, instead of recording every instantaneous value of the wave, the sine wave is sampled and recorded at uniformly spaced discrete intervals; the original sine wave being reconstructed on playback by passing the discontinuous readings through an appropriate filter.

If a signal is being sampled at discrete intervals it is possible to use the time between them for sampling other signals (see page 279). This may be accomplished by using some form of scanning switch or multiplexer[10,11]. A schematic diagram of such a system is shown in figure 9.7, where if a total of 900 samples are made via a 90 position scanner, 86 channels of input can be recorded per tape track, providing that the frequencies of the input frequencies are all less than 1.5 Hz. The four scanner positions not connected to input channels being used for zero and maximum voltage calibration levels and synchronising purposes.

Thus the chief advantage if the p.d.m. recording process is its ability to record 'simultaneously' a large number of channels of information. Other advantages are its high accuracy, due to the possibility of selfcalibration, and a high signal to noise ratio. The disadvantages are the very limited frequency responses and the complexity of the auxiliary electronic equipment.

The direct and f.m. processes are those generally used in instrumentation, a number of commercial recorders being available with input modules designed so that the signal may be recorded by either direct or f.m. recording, depending on which is the more suitable for the particular signal. The p.d.m. process is normally only used for special applications such as flight recorders, where a large number of slowly changing variables are to be recorded. Digital tape recording is rarely used in instrumentation (apart from storing the output from data loggers), its major application being associated with digital computing and control.

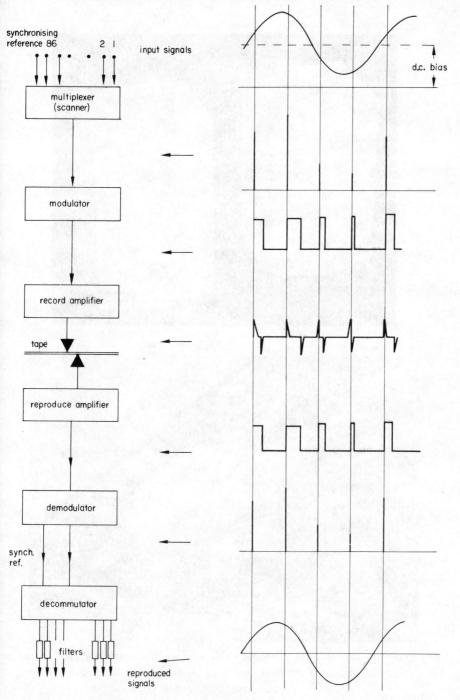

Figure 9.7. Schematic diagram of pulse duration modulation recording system

(a)

(b)

Figure 9.8. (a) A 14 track interlaced IRIG head set (Courtesy Gresham Recording
Heads Ltd); (b) Dual head stack mounting in an Ana-log 7 recorder.
The transparent tape is part of a 'lead in' system for automatic
threading of the magnetic tape (Courtesy Pye–Unicam/Philips)

9.2.4. Magnetic Heads

The limitations of each recording process are closely related to the specific characteristics of the record and reproduce heads. Each head consists or two identical core halves built up of thin laminations of a material that has a high magnetic permeability, and good wear properties, to ensure high efficiency, low eddy current losses, and a long life. The two halves of the core are wound with an identical number of turns and assembled with nonmagnetic inserts to form the front and back gaps. Only the front gap contacts the tape and enters into the recording or reproducing process, the presence of the back gap is purely to maintain symmetry and minimise the effects of random pickup in the windings. In a record head the gap must be large enough to achieve deep flux penetration into the tape and short enough to obtain sharp gradients of high frequency flux at the trailing edge, for a particle of magnetic oxide crossing the gap remains in a state of permanent magnetisation proportional to the flux that is flowing through the head at the instant that the particle passes out of the gap. The gap length (typically $5-15$ μm) of a record head has little effect on the instrument's frequency response, but the accuracy of the trailing gap edges is of extreme importance and considerable care is taken in its manufacture to obtain a sharp well defined gap edge. In a reproduce head, the front gap size ($0.5-5$ μm) is a compromise between upper frequency limit, dynamic range, and head life.

In multitrack heads (see figure 9.8a and b) a compromise must be established between track widths (wide tracks giving a good signal to noise ratio), track spacing (large spacing preventing crosstalk), and overall width of the tape which it is desirable to keep to a minimum. Typical dimensions are: track width, 0.5 mm, and spacing between tracks 0.75 mm, giving four tracks on a 6.25 mm wide tape. The small spacing between tracks means that to reduce the noise level between the heads of a stack, an electromagnetic screen must be positioned between them[12], as well as using two heads offset as shown in figure 9.9. In a multitrack head it is essential that close tolerances are maintained on mechanical positioning and in electrical characteristics so that uniformity of response from track to track and from one recorder to another is maintained. It should be noted that for digital recorders the signal to noise requirements are less severe and as many as 16 tracks may be used on a 25 mm tape.

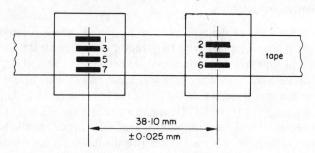

Figure 9.9. Offset of multitrack heads

9.2.5. Tape Transport

To move the tape past the heads at a constant speed a 'tape transport' is used. This must also be capable of handling the tape during the various modes of operation without straining, distorting or wearing the tape. To accomplish this, a tape transport must guide the tape past the heads with extreme precision and maintain the proper tension within the head area to obtain adequate tape to head contact. Spooling or reeling of tape must be performed smoothly so that a minimum of perturbations are reflected into the head area. Take-up torque must be controlled so that a good pack results on the take-up reel. It is also the job of the transport to move the tape from one reel to the other quickly when rewinding. Even with fast speeds, the tape must be handled gently and accurately so that a good tape pack is maintained on each reel. In going from a fast speed to stop (or vice versa) precise control of the tape must

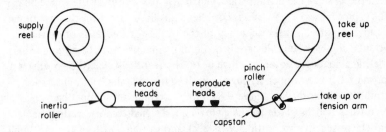

Figure 9.10. Simple tape transport system

be maintained so that undue stress or slack is not incurred in the tape. A simplified arrangement of a tape transport is shown in figure 9.10 where a capstan and pinch roller are used to drive the tape[9,10,11]. Some manufacturers use a closed loop drive, and others a two capstan arrangement. Considerable development work being in progress to establish even better tape transport systems.

9.2.6. Tape Motion Irregularities

Variations in the motion of the tape give rise to: *flutter,* which is caused by variations in the tape speed that have a frequency greater than 10 Hz; *wow,* which is caused by tape speed variations that are between 0.1 and 10 Hz; and *drift,* which denotes variations in the tape speed accuracy that are below 0.1 Hz. Common usage has broadened the definition of the term *flutter* to include all variations in the range 0.1 Hz to 10 kHz and in a recorder specification this should be quoted as either a r.m.s. or peak to peak value over a specific band of frequencies. The term *skew* is used to describe the fixed and variable time differences associated with a particular head stack. It implies that the tape is

not moving in a longitudinal manner as it passes the heads. Fixed or static
skew contributes a constant relative timing difference, and dynamic skew
produces a variable timing difference between tracks.

Fixed skew is usually caused by the misalignment of a head to the tape,
misalignment of individual heads in a stack, and bad guiding that produces
differences in tension distribution as the tape crosses the heads. Dynamic skew is
produced when there is a varying uneven tension distribution across the tape,
which may be initiated by tape scrape, vibration, and tape irregularities. Fixed
skew and some forms of dynamic skew produce relative timing errors between
tracks proportional to the spacing between tracks. Thus tape transport specifications usually show skew (static and dynamic) or total interchannel displacement
error as a time interval for a particular tape speed, for example ± 13 pm: 0.5 μs
at 3.05 m/s.

Flutter and skew errors are imposed by both the recording and reproducing
operations, and tapes reproduced on the same machine as they were recorded
on will have less skew error than tapes reproduced on different machines.

Stretching of the tape causes a timing error between odd and even tracks of
the tape. It was indicated in figure 9.9 there is a separation of 3.8 cm between
the two head stacks[12], and if the tape changes length between recording and
reproducing due to stretching or environmental conditions, a timing error will
be created between odd and even tracks.

Features

The advantages of the tape recorder, compared with the data logger, are that
it facilitates a continuous record of a number of signals, which may have a wide
range of frequencies, to be made simultaneously. This has distinct advantages in
the study of transient and 'once only' situations, for example, car crashing at
road research establishments. When using a tape recorder it is possible to 'play
back' a recording any number of times without deterioration of the record: it
is also possible to change the time scale between recording and reproducing a
signal, that is, a 'fast' occurrence may be recorded with a fast tape speed
(1.52 or 3.05 m/s) but for analysis of the signals the tape could be played back at
a slow speed 4.76 or 2.38 cm/s), enabling the recorded signal to be traced on
an X–Y plotter (see page 47). The reverse time scaling procedure is also possible.
that is a slow process could be recorded with a tape speed of (say) 2.38 cm/s and
reproduced from a tape speed of 1.52 m/s.

Some instrumentation tape recorders have 'level' sensing circuits on the
input so that if the instrument is being used for data recording and a preset limit
on the magnitude of the input signal is exceeded, an alarm indicator may be
operated.

The main disadvantage of the instrumentation tape recorder is that the
recorded signals on the tape are in no way visual, and must be reproduced before
any analysis may be performed.

9.3 DATA ANALYSERS

Automatic data logging and recording result in large quantities of data that must be analysed. This may be accomplished by feeding the data into a digital computer that has been suitably programmed, or by using an instrument that has been specifically designed for coping with the analysis of a particular type of data.

9.3.1. Pulse height analysers

For large quantities of data to have a meaningful form it is usually necessary to adopt a graphical presentation, for example, a histogram of a large quantity of data is easier to assess than columns of figures. To perform this type of analysis, instruments have been manufactured that can sense the magnitude of a pulse and · classify it as belonging to one of a number of predetermined bands of pulse height, it then being possible to display the histogram on a c.r.t. or, in a slow speed statistical analyser, as a count of pulses within the preset bands of pulse height[13].

Such a device is suitable for the analysis of statistical data, for example the weight of potatoes, the radiation levels from an isotope, etc., but much of the time spent in analysing data is consumed in trying to form relationships between the occurrence of events, and to locate the source of a phenomena; in other words to correlate the effect with a cause, when the signals from both may be partially or completely obscured by noise.

9.3.2. Correlator

The development of the signal correlator has reduced the correlation of signals from an extremely time consuming two stage process (data recording and computer analysis) to a 'real time' process. The simplest method of obtaining a factor that describes the correlation between two waveforms is to multiply sampled amplitudes of the two waveforms at regular intervals throughout the

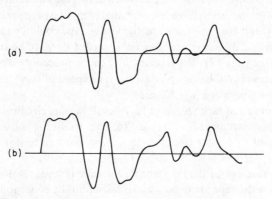

Figure 9.11. Waveforms with good correlation

duration of the waveform. If the waveforms to be compared (see figure 9.11) are identical and in phase (no time displacement) then the summation of the products obtained will result in a large positive number. However, should the waveforms have been different the final number would have been smaller, due to some products having a negative value. If the waveforms to be compared are identical but with a time displacement (see figure 9.12), then a direct comparison would yield a low correlation, but by varying the time shift (τ) between the two waveforms it is possible to obtain a curve of average correlation product against time shift.

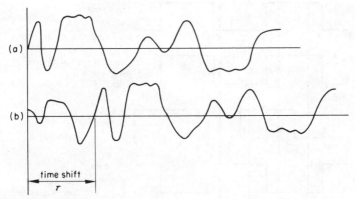

Figure 9.12. Waveforms that have good correlation when one is time shifted by τ

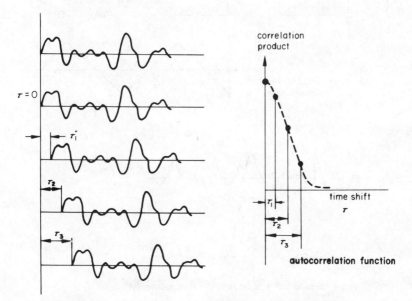

Figure 9.13. Autocorrelation or variation of correlation with time

Autocorrelation Function

The above results in a curve known as the autocorrelation function (see figure 9.13). The average correlation product for each value of τ is determined by dividing the total of the correlation products for each value of τ by the number of products contributing towards it. Thus the autocorrelation function of a waveform is a graph of the similarity between a waveform and a time shifted version of itself, as a function of the time shift. The autocorrelation function of

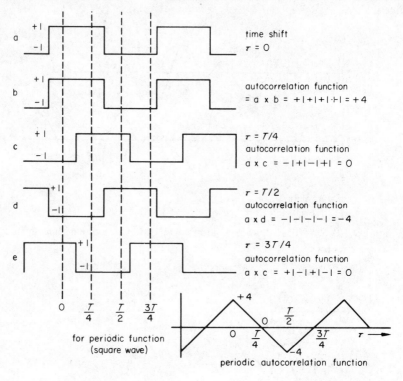

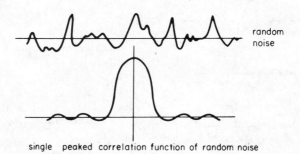

Figure 9.14. Autocorrelation functions of periodic and nonperiodic waveforms

any signal, random or periodic, depends not on the actual waveform but on its frequency content. This means that the autocorrelation function of any periodic wave is periodic and of the same period as the waveform, while the autocorrelation function of a wide band nonperiodic waveform will be nonperiodic (see figure 9.14). Autocorrelation is uniquely successful in the determination of the periodic content of an unknown noisy signal, a striking example of this being the detection of periodic signals from outer space, for example emissions from pulsars.

Cross-Correlation Function

To compare the similarity of two nonidentical waveforms a technique similar to that which gave the autocorrelation function may be used to obtain a cross-correlation function. An example of this is in correlating a signal with a reflection that has been modified by noise. Now in general there will be no correlation between the original signal and the noise (giving a high signal to noise ratio), and the cross-correlation function between the original signal and the reflection plus noise will take the form of an autocorrelation function having a delay that is proportional to the transmission time between sending the signal and receiving the reflection. Figure 9.15 shows (a) a swept frequency transmitted signal, (b) a received signal that has a noise content, and (c) the cross-correlation function of the two signals, which is a delayed version of the autocorrelation function of the transmitted waveform.

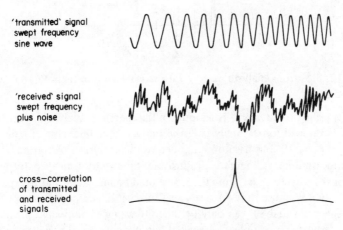

'transmitted' signal
swept frequency
sine wave

'received' signal
swept frequency
plus noise

cross—correlation
of transmitted
and received
signals

Figure 9.15. Detection of a known signal 'buried' in noise

It may be shown[14,15,16] that the relationship between the autocorrelation function of a signal and its power density spectrum is a Fourier transform pair. This enables the autocorrelation function to be related to the power density (spectrum) as measured with a wave analyser fitted with a true square law meter.

9.3.3. Spectrum analyser

In communication systems extensive use is made of modulated carrier frequencies to convey data, speech, and other signals. The display on a conventional oscilloscope of even a simple modulation of a carrier wave is difficult to analyse in respect of the magnitudes of the various frequency components[17], for example, see figure 9.16. If, however, the modulated waveform could be displayed as a function of frequency instead of time, the relative magnitudes of the frequency components would become apparent. This form of display — magnitudes of

Figure 9.16. Oscilloscope display of a 1 MHz waveform modulated by a 1 kHz waveform

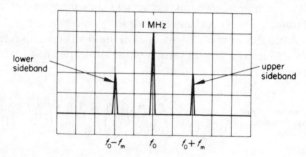

Figure 9.17. Spectrum analyser display for the waveform in figure 9.16 (Tektronix)

frequency against frequency — is adopted in the spectrum analyser, which was originally developed for the analysis of the frequency components of radio frequency signals[18,19], but has now extended to the analysis of vibration and accoustic measurements. Figure 9.17 illustrates the spectrum analyser form of display for the waveform in figure 9.16. It should be indicated that the spectrum analyser is limited in use to repetitive waveforms, since at any one instant of time the spectrum analyser can only be tuned to a single component of the frequency spectrum, and unless a transient component coincides with its value in the sweep of frequency it will not be displayed.

9.4 COMPARISONS

In table 9.1 is presented a comparison table of the various forms of recording instrument. For completeness the graphical recorders of chapter 2 are included.

Table 9.1 Comparison Table of Data Recording Instruments

	Type	Typical Errors	Maximum sensitivity	Input Impedance	Frequency Range	Record	Remarks
	(d. vm.) Date logger	±0.1 → 1%	10 μV	10 MΩ	d.c. → 1 Hz max. (depends on scanning and reading rates)	Printed or punched tape	20–1000 channels Sampled
	(Counter) Data logger	±0.1 → 1%	±1 count	10 kΩ → 1 MΩ	ditto	ditto	ditto
Tape Recorders	Direct Record	±1 → 10%	100 μV	100 kΩ	50 Hz → 2 MHz	stored on magnetic tape, play back to get output on graphic recorder	4–18 channels may change time scale between record and reproduce, p.d.m. may have 900 channels
	f.m.	±0.1 → 1%	100 μV	100 kΩ	d.c. → 50 kHz		
	p.d.m.	±0.1 → 1%	100 μV	100 kΩ	d.c. → 0.01 Hz		
Moving Coil	ink pen	±1%	4 mV/cm	100 Ω	d.c. → 5 Hz	graph of variable against time	single channel
	chopper bar	±1%	4 mV/cm	>100 Ω	d.c. → 0.01 Hz		
	heated stylus	±1%	4 mV/cm	>100 Ω	d.c. → 20 Hz		
	electrostatic	±1%	100 mV/cm	20 kΩ	d.c. → 60 Hz		
	Potentiometric ink pen	±0.5%	4 μV/cm	>100 kΩ	d.c. → 8 Hz	ditto	1–24 channels linear scale 1 channel circular scale
	u.v./light spot recorder	±5% ±5% ±5%	0.65 μA/cm 4.73 mA/cm 18.2 mA/cm	130 Ω 69 Ω 57 Ω	d.c. → 6 Hz d.c. → 400 Hz d.c. → 5000 Hz	ditto ditto ditto	1–50 channels record has limited life unless permanised
	c.r.o.	±5%	10 μV/cm	10 MΩ	d.c. → 50 GHz	ditto	visual or photographic
	Transient recorder	±5%	10 mVf.s.	1 MΩ	time interval 200 μs to 20 s	stored in digital form	requires display instrument

REFERENCES

1   G. A. Rigby. 'Data Logging'. *Control and Instrumentation*, (Nov. 1970)
2   E. J. Wightman. 'Design Fundamentals of Data Logging'. *Instrument Practice*, (April 1967)
3   J. E. Freebody. *Telegraphy*, Pitman, London (1958)
4   P. D. Gotting. 'Data Logging in the 70s'. *Control and Instrumentation* (May 1970)
5   R. F. Chapman, 'Industrial Data Acquisition for the 70s'. *Electronics and Power* (May 1970)
6   S. Stein and J. Jay Jones. *Modern Communication Principles*, McGraw-Hill, New York (1967)
7   R. H. Cerni and L. E. Foster. *Instrumentation in Engineering Measurement*, Wiley (1962)
8   E. A. Read. 'Analogue Recording'. *Electronic Equipment News* (Oct. 1970)
9   Magnetic Tape Recording Handbook. *Hewlett Packard Application Note 89.*
10 C. B. Pears Jnr (Ed.) *Magnetic recording in science and industry*, Reinhold (1967)
11 G. L. Davis. *Magnetic Tape Instrumentation*, McGraw-Hill, New York (1961)
12 IRIG Telemetry Standards, document no. 106–69. *Defence Documentation Centre for Scientific and Technical Information.* (Ministry of Technology, Orpington, Kent)
13 Statistical Analysis of Waveforms and Digital Time-waveform Measurements. *Hewlett Packard Application Note 93.*
14 R. L. Rex.and G. T. Roberts. 'Correlation, Signal Averaging and Probability Analysis'. *Hewlett Packard Journal* (Nov. 1969)
15 F. H. Lange. *Correlation Techniques*, Iliffe, London (1967)
16 Jens Trampe Broch. 'On the Measurement and Interpolation of Cross-power-spectra'. *Brüel and Kjaer, Technical Review No. 3* (1968)
17 R. Myer. *Getting Acquainted with Spectrum Analysers.* Tektronix Inc. (1965)
18 D. Welch. 'Spectrum Analyser Measurements'. *Tektronix Measurement Concepts* (1969)
19 J. Søeberg. 'Real Time Analysis'. *Bruel and Kjaer Technical Review No. 4* (1969)

10

Instrument Selection and Specifications

The considerations for selecting an instrument may be regarded as falling into two categories: either an engineer is selecting the most suitable instrument from those within a department or establishment to perform a particular measurement, or he is undertaking the purchase of a new instrument to perform a particular measurement and possibly at the same time extend the measurement capabilities of the department or establishment in which he works. Many of the criteria in selecting an instrument are the same, whether the engineer is selecting an instrument off the shelf or purchasing new equipment. In either case a major pitfall is to 'acquire' the newest and most sophisticated pieces of equipment in the department, or on the market, simply as a prestige exercise. This is of little value if within a week, justifiable pressure is brought to bear by one's colleagues and the prestige instrumentation system is reduced to its minimum requirements – which could be two suspect multimeters and the oldest 'scope in the department!

10.1 INSTRUMENT SELECTION

The general criteria for selecting an instrument[1,2,6] may be summarised by the following check list, which although it may be more suitable when considering a moderately sophisticated instrument, could prove valuable as a guide in selecting the 'right' instrument on every occasion.

Ranges
    (a)   What are the maximum and minimum magnitudes of the values to be measured?
    (b)   Will a single range, or multirange instrument be the most suitable?

Accuracy
    (a)   What is the maximum tolerance acceptable?
    (b)   Is the resolution of the instrument consistent with its specified errors

Response characteristics

   (a)   What is the acceptable response time/bandwidth?

   (b)   Are frequency compensating probes required?

   (c)   For autorange (for example, d. vm.) instruments, must (a) include the time for range and polarity changes?

   (d)   For a.c. instruments is it desired to sense mean, peak, or r.m.s. values?

Input characteristics

   (a)   What are the allowable limits of instrument input impedance?

   (b)   Can calculations be made to correct for the instrument loading on the measurand?

   (c)   Is the instrument input impedance constant for all ranges? If not are the variations in magnitude acceptable?

   (d)   Are there any source impedance restrictions? If so will they affect the operation of the instrument?

Output characteristics

   (a)   What form of display is required? For example, graphical, digital, etc.

   (b)   Is an electrical output required to operate in conjunction with other equipment? If so, what signal levels are required, and what codes are used?

Stability

   (a)   What is the maximum acceptable time between calibrations?

   (b)   Is the instrument to be operated unattended for long periods?

   (c)   Is there a 'built in' calibration system?

Environment

   (a)   Over what range of temperature, humidity, line voltage variations, etc., will the instrument be required to operate, and how do these factors affect the errors?

   (b)   Will the instrument be subjected to mechanical shock or vibration? If so, what will the fundamental frequency be?

   (c)   Are there any size limitations?

   (d)   If the instrument is to be permanently installed will access be required for maintenance?

Isolation and screening

   (a)   Will the instrument be subjected to stray electromagnetic or electrostatic fields?

   (b)   Is the measurand 'floating' or has it one side earthed?

(c)   Is battery operation or 'guarding' required to ensure adequate performance?

(d)   What is the rejection of d.c. and a.c. common mode voltages?

Operation

(a)   Is remote control required?

(b)   Is automatic or programmed operation required?

(c)   Will multifunction operation be advantageous?

(d)   Will operator fatigue cause reading interpretation problems?

(e)   What power supplies will be required?

Reliability

(a)   What is the specified operational life?

(b)   What will the consequences of failure be?

(c)   Is duplication or standby instrumentation required?

(d)   Will special spares and maintenance equipment be required?

(e)   Does the instrument incorporate any limit detection or alarm facilities? If so, must a 'fail safe' arrangement be incorporated?

On completion of the above check list for a particular application, the derived specification for the desired instrument may not be possible in practical terms, and a compromise between that which is available within an organisation, or can be afforded, will have to be adopted.

If a new instrument is to be purchased it is essential to ensure that the 'right' instrument is being purchased. This is particularly relevant if the instrument is for a permanent installation although it may be considered as 'good practice' to purchase to a slightly higher specification if the proposed instrument is for use in a laboratory where the measurement requirements may change with experience and time. The problem here, of course, is overspecification resulting in the purchase of an instrument that is much more sophisticated (and expensive) than is really necessary.

The following list is a rough guide to the factors that will increase the cost of an instrument:

(a)   reduction of error magnitudes

(b)   increasing the speed of reading

(c)   increasing the sensitivity

(d)   improving the stability

(e)   improvement of isolation/guarding (except for battery operation)

(f)   the addition of output features

(g)   the extension of operating conditions.

Thus having decided on the requirements of the instrument it is desirable to purchase, it is necessary to study manufacturer's literature and decide on a 'best buy'.

10.2 SPECIFICATION ANALYSIS

Since each manufacturer will write his specification sheet for a particular instrument so as to emphasis what he considers to be his product's most important merits, the emphasised features for a particular type of instrument may be different for each of a number of competitive units. It is therefore necessary to extract from the specification sheets the data relevant to one's specific requirements and reduce this to a common level for comparison purposes.

Probably one of the greatest uses of 'spec-manship' is in the specification of the errors of digital voltmeters, which are normally quoted in two parts, for example:

(a)   0.01 per cent of reading ± 0.01 per cent of full scale; or
(b)   0.01 per cent of reading ± 1 digit.

Taking these two forms of specification (and there are others), let us compare the maximum uncertainty in instrument reading that is possible if the two forms of specification are applied firstly to an instrument that has a full scale of 9999; and secondly to an instrument that has a full scale of 4999.

For the 9999 full scale instrument.

When reading full scale, the maximum possible uncertainty in reading is 0.01 per cent + 0.01 per cent = 0.02 per cent of reading, and over a range will vary as shown in table 10.1.

Table 10.1

Reading	% f.s.	Quoted % uncertainty + in reading	Uncertainty quoted as a % of f.s. as a % of reading	=	Total uncertainty as a % of reading
9999	100	0.01	$\left\{\dfrac{0.01}{100} \times \dfrac{9999}{9999} \times 100 = 0.01\right\}$		0.02
5000	50	0.01	$\left\{\dfrac{0.01}{100} \times \dfrac{9999}{5000} \times 100 = 0.02\right\}$		0.03
2000	20	0.01	$\left\{\dfrac{0.01}{100} \times \dfrac{9999}{2000} \times 100 = 0.05\right\}$		0.06
1000	10	0.01	$\left\{\dfrac{0.01}{100} \times \dfrac{9999}{1000} \times 100 = 0.10\right\}$		0.11

Since 1 digit = $100 \times 1/9999$ = 0.01 per cent of full scale the two forms of specification are equivalent for this instrument.

Consider now the 4999 digital voltmeter:

Here 1 digit = $100 \times 1/4999$ = 0.02 per cent of full scale and an accuracy quoted as ± 0.01 per cent of reading ± 0.01 per cent of range would be incompatible with the resolution of the instrument, for there will always be an uncertainty of at least ± 1 digit in the display of a digital instrument. Thus the error specification for the 4999 d. vm. must be of the form ± (0.01 per cent of reading + 1 digit) and the maximum possible uncertainty in the reading over a range may be tabulated in table 10.2.

Table 10.2

Reading	% f.s.	Quoted % uncertainty + in reading	Uncertainty quoted as 1 digit, as a % of reading	=	Total uncertainty as a % of reading
4999	100	0.01	$\frac{1}{4999} \times 100 = 0.02$		0.03
2000	40	0.01	$\frac{1}{2000} \times 100 = 0.05$		0.06
1000	20	0.01	$\frac{1}{1000} \times 100 = 0.10$		0.11
500	10	0.01	$\frac{1}{500} \times 100 = 0.20$		0.21

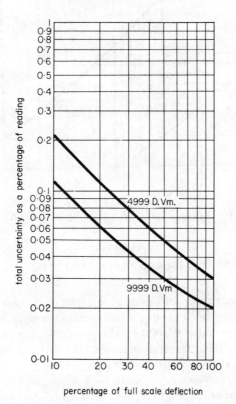

percentage of full scale deflection

Figure 10.1 Errors plotted against percentage of full scale

The values of uncertainty tabulated above may, for comparison purposes, be presented in graphical form. By using horizontal axes of (a) percentage of full scale, and (b) reading in volts, two different pictures are presented (see figure 10.1 and 10.2). The latter graph shows that in measuring voltages between 1.000 and 5.000 there is no difference in the magnitude of the uncertainty in the reading when measured by either instrument, whereas this is not at all apparent from the curve in figure 10.1.

The question of digital voltmeter error specification may be further obscured by the influence of temperature. For example the above form of

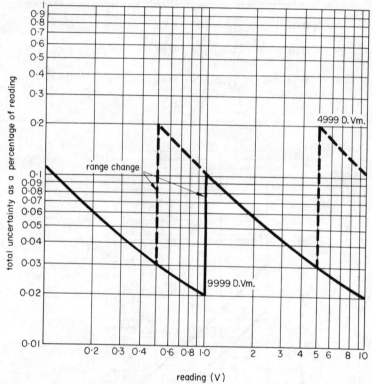

Figure 10.2. Errors plotted against reading

error specification may apply for a range of temperatures from 15 to 40°C for a particular instrument, whereas other instruments of the same type may have temperature coefficients that affect the accuracy specification. A typical method of specifying this effect is to quote the temperature coefficient as a percentage of reading plus a percentage of range.

For example, ±(0.005 per cent of reading + 0.002 per cent of range) per °C and such effects must be taken into account when the errors of two or more instruments are being compared.

As an example, let us reconsider the 9999 d. vm. and 4999 d. vm. compared above, assuming that the full error specification of the 9999 instrument is

± (0.01 per cent of reading + 0.01 per cent of range) at 20°C
± (0.005 per cent of reading + 0.002 per cent of range) per °C

and for the 4999 instrument is

±(0.01 per cent of reading + 1 digit) over a temperature range 10 to 40°C.

Let it further be considered that the digital voltmeter purchased will be required to operate in an environment that has a temperature range from 20 to 30°C. The maximum uncertainty in the reading on the 4999 instrument will be as calculated before but an uncertainty of + (0.005 per cent of reading + 0.002 per cent of range) × 10°C/°C must be added to the values previously tabulated for the 9999 d. vm. so that an allowance is made for the worst possible uncertainty, that is, that which will occur at 30°C, as shown in table 10.3.

Thus the error comparison curves for our hypothetical digital voltmeters at 30°C are as shown in figure 10.3 and this presents a very different picture to that indicated by figure 10.1. This should illustrate that considerable thought must be given before arriving at a decision on the most suitable of a number of instruments, particularly if the ranges are overlapping, for example 1, 10, 100, 1000 V for the 9999 instrument   and 0.5, 5, 50, 500, 1000 V for the 4999 instrument.

Table 10.3

Reading	% uncertainty at 20°C	Effect of temperature coefficient		Total uncertainty at 30°C
		portion quoted as % of reading	portion quoted as % of range	
9999	0.02	0.05	0.02	0.09
5000	0.03	0.05	0.04	0.12
2000	0.06	0.05	0.10	0.21
1000	0.11	0.05	0.20	0.36

It must be emphasised that while accuracy is of major importance, it is not the only consideration when selecting an instrument, it being equally important to compare all the facets of the instruments under consideration.

Consider the type of exercise that an engineer should conduct if desiring to purchase an instrument capable of measuring d.c. voltages between 100 mV and 500 V with an error less than 0.05 per cent. There is a wide range of instruments available for this purpose which may be divided into three groups, namely digital voltmeters; differential voltmeters; and d.c. potentiometers with their accessories. The details of an instrument typical of each of the three groups and available in 1970 are compared in table 10.4, the particular instruments detailed having been selected at random from a range of competitive instruments in each group.

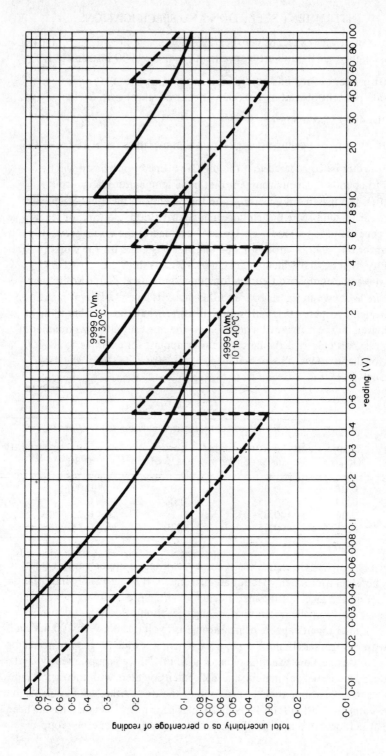

Figure 10.3. Error comparisons at 30°C for the hypothetical d. vms

From table 10.4 the d.c. potentiometer would appear to have advantages in that it has the greatest number of ranges, and the best resolution, that is 5 μV. The d.c. potentiometer is at a disadvantage if voltages greater than 1.89 V are to be measured since such values require the use of the voltage ratio box which has an input impedance of 100 Ω/V, compared with a minimum of 10 000 Ω/V input impedance of both the differential and digital voltmeters. Ambient temperature variations would appear to have the least effects on the performance of the differential voltmeter, although at the extremes of temperature the temperature coefficient associated with the digital voltmeter would have the smallest effect on the uncertainty in the instrument's reading.

Of the instruments compared the only one supplied with an output is the differential voltmeter, which has an output suitable for linking to a pen recorder. The digital voltmeter of the type considered may be fitted with a binary coded output at additional expense. The major advantage of the digital voltmeter is the speed at which an unknown may be measured, plus the fact that its operation requires only the minimum of effort from the operator, that is that the operator selects the correct range on which to measure the unknown.

To compare the error specifications of the three instruments over the range 100 mV–500 V the curves in figure 10.4 have been drawn of the maximum uncertainty in reading against voltage for operation at a temperature of 20°C. These curves indicate that the differential voltmeter has the best accuracy performance above 100 mV, while between 18 and 100 mV the digital voltmeter is the best, and below 18 mV the potentiometer has the least uncertainty in its reading. It must be remembered that these error curves are for the conditions stated on the specification and that only in the case of the differential voltmeter is a temperature range quoted. Thus to perform a thorough analysis an allowance should be made for the effects of temperature on the total uncertainty in a reading. It was assumed that the specifications for the potentiometer and the digital voltmeter were given for operation at 20°C, and if the voltmeter that is brought will be operated within reasonable variations of ambient temperature, ·say, 15–25°C. Then an additional 0.02 per cent of reading should be added to the uncertainty when measuring any value with the potentiometer; while the uncertainty for all values of digital voltmeter reading is increased by 0.01 per cent of the reading; the curve for the differential voltmeter reading remaining unaltered. Thus it would appear that from purely an accuracy consideration the differential voltmeter is the 'best buy'.

It is important to emphasise however that the other aspects listed must be carefully considered, and that before a final decision is made, arrangements should be made for the manufacturers of the various instruments to demonstrate their instruments under conditions in which they will be required to operate.

Table 10.4

Instrument[3,4,5]	Cropico d.c. potentiometer and accessories	J. Fluke Differential voltmeter	S.E. Labs. Digital voltmeter
Type No.	P.3/V.R.3/P.3/s	871 A	SM 210
Ranges	(a) 0–0.018 V (b) 0–0.18 V (c) 0–1.8 V (d) also 15, 30, 70, 150, 300, 600, 1800, 2400 V using range box	(a) 0–1.09999 V (b) 0–10.9999 V (c) 0–109.999 V (d) 0–1099.99 V	(a) 0–99.99 mV (100 mV) (b) 0–0.9999 V (1 V) (c) 0–0.000 V (10 V) (d) 0–00.00 V (100 V) (e) 0–999.9 V (1000 V)
Resolution (on ranges a, b, c, d, and e)	(a) 5 μV (b) 50 μV (c) 500 μV (d) 0.028% of range box tapping	(a) 10 μV (b) 100 μV (c) 1 mV (d) 10 mV	(a) 10 μV (b) 100 μV (c) 1 mV (d) 10 mV (e) 100 mV
Display	'dialled' digits full scale 1.8995	in line digits + decimal point full scale X 9999	numerical indicator tubes + decimal point full scale 9999
Input Impedance (on ranges a, b, c, d, and e)	(a) ⎫ (b) ⎬ tending to ∞ at balance (c) ⎭ (d) 100 Ω/V	(a) ⎫ tending to ∞ at balance (b) ⎭ (c) > 10 MΩ that is at (d) least 10 000 Ω/V	(a) ⎫ (b) > 1000 MΩ, that is at (c) least 100 MΩ/V (d) ⎫ > 10 MΩ, that is at (e) ⎭ least 10 000 Ω/V
Accuracy (on ranges a, b, c, d, and e)	(a) 0.04% of reading or 3 μV ⎫ (b) 0.03% of reading or 25 μV ⎬ whichever is greater (c) 0.02% of reading or 200 μV ⎭ (d) as (c) plus 0.02% for range box	All ranges ± (0.02% of reading + 0.001% of range)	(a) ⎫ (b) ⎬ ± 0.01% of range ± 1 digit (c) ⎭ (d) ⎫ 0.015% of range ± 1 digit (e) ⎭

Operating temperature range	Not quoted on spec.*	within spec. 10–40°C 0–10°C and 40–50°C ± (0.05% of input voltage + 10 µV)	0–+55°C Temp. coefficient ± 20 p.p.m./°C over operating temp. range
Voltage ref.	Standard cell (stability not quoted on specification but given seperately as 10 p.p.m./yr)	Temp. comp. zener diode stability within ± 0.002% per hand ± 50 p.p.m./yr. Temp. coefficient less than ± 5 p.p.m./°C	Unsaturated Weston cell. Temp. coefficient ± 5 p.p.m./°C
C.M.R. d.c.	Not quoted, but should be high	> 130 dB	> 100 dB
Output	None	0–±20 mV linked to meter display	None on SM 210, other models with B.C.D. output at additional expense
Reading rate	Depends on operator, 30–60 s	Depends on operator 20–40 s	2 per second
Price January 1971	P.3 – £141 VR.3 – £120 P.3/s – £20 Total: £281	£282	£295

* The standard cell voltage is compared with the volt drop across a fixed value resistor that also has a negative temperature coefficient so that the nett temperature effect is approximately zero.

Note. Acknowledgement is made to the manufacturers for permission to reproduce these particulars.

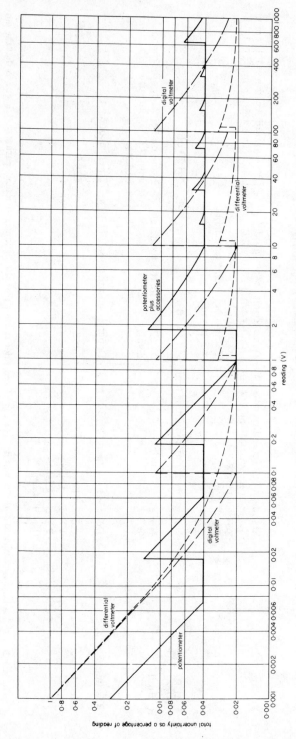

Figure 10.4. Error comparison at 20°C for the three instruments considered

Table 10.5 Instrument selection

	voltage d.c.	voltage a.c.	current d.c.	current a.c.	power	resistance	inductance	capacitance	frequency	phase angle	comp. table page no.
pointer instruments — moving coil	√		√								
moving iron	√	√	√	√						√	
electrodynamic	√	√	√	√	√					√	
rectifier		√		√							
thermocouple	√	√	√	√	√						41
electrostatic	√	√			√						
electronic d.c. — direct coupled	√		†								
electronic d.c. — chopper amplifier	√		†								
electronic a.c. — mean sensing		√		√							
electronic a.c. — peak sensing		√		√							
electronic a.c. — r.m.s. sensing		√		√							
'Q' meter						*	√	√		*	115
graphical instruments — moving coil — ink pen	√	√	√	√							
moving coil — chopper bar	√		√								
moving coil — heated stylus	√	√	√	√							74, 294
moving coil — electrostatic stylus	√	√	√	√							
potentiometric	√	√	†	†							
u.v./light spot	√	√	√	√							
c.r.o.	√	√	†	†							
x–y plotter	√	√	†	†							
d.c. potentiometer	√		†		†	†					
a.c. potentiometer		√	†	†		†	†	†		†	
wheatstone bridge						√					115
kelvin double bridge						√					
a.c. bridge circuits						√	√	√	√	*	
transformer ratio		√		†		√	√	√		√	
double ratio bridge						√	√	√		√	
twin 'T'						√	√	√	√		109
differential voltmeter	√	√	†	†		√ †					
counter timer									√		206
digital voltmeter	√		†								

† = by volt drop

* = leakage resistance, giving loss angle

10.3 INSTRUMENT COMPARISONS

A large number of instruments have been described, and since in some measurement situations more than one instrument may be suitable while in others none of the available instruments are appropriate, generally the first question that is asked is; which instrument is the best to use? To assist in the solution of this problem the selection table 10.5 has been drawn up, and using this together with the comparison tables at the ends of the earlier chapters, and the selection factors listed in this chapter it should be possible to select the optimum instrument for most measurement situations. It must, however, be emphasised that no attempt has been made in this book to detail the particular instruments of the specialist, for these are not of general interest and are best described by specialists in the appropriate subject.

REFERENCE

1 *Hewlett Packard Application Note 69* (1965)
2 *Hewlett Packard Application Note 60* (1965)
3 *Leaflet No. 54.* Croydon Precision Instrument Co. (June 1969)
4 *Specification Digital Voltmeters SM210, SM210/A and SM210/B.*
      SE Laboratories (Engineering) Ltd.
5 *871 A Solid-state D.C. Differential Voltmeter.* John Fluke Mfg. Co. Inc.
6 *Instruments in Working Environments. Sira Conference, May 1970.*
      Adam Hilger Ltd (1971).

11

Instrumentation Systems

Definition

Instrument systems refine, extend, or supplement human facilities and abilities to sense, perceive, communicate, remember, calculate or reason[1,8].

To relate this definition to practical terms means that any use of instruments constitutes an instrumentation system, since a suitable instrument or chain of instruments will always convert an unknown quantity into a record or display that human facilities can interpret.

11.1 OPEN AND CLOSED LOOP SYSTEMS

The basic form of an instrumentation system consists of the measurand, a transducer or sensor, signal conditioning equipment, and the display or recording instrument (see figure 11.1). Such a system simply converts the measurand into a record or display so that an operator may observe the variations in, or magnitude of, an unknown. This type of system becomes more complex as sophisticated automatic recording is added, enabling an increase in the number of measurands that may be observed and recorded by a single operator. The addition of such equipment may also enable simple computations to be performed on the quantities measured.

Figure 11.1. Basic open loop instrumentation system

If an open loop instrumentation system is used in a manufacturing process it will require the continual vigilance of an operator to maintain the measurand (for example, temperature of a furnace) at a predetermined level. However, by adding a reference, some form of comparator, and a feedback path to the basic system a closed loop system is formed (see figure 11.2). This latter type of system

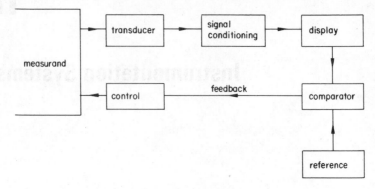

Figure 11.2. Closed loop instrumentation and control system

is finding an ever increasing number of applications in manufacturing industries where automation increases the speed of production and reduces the wages bill. It is also the basis of control engineering, which is a continually developing branch of engineering[2].

11.2 BENCH INSTRUMENTATION

The improvement of specifications and increase in complexity of manufactured equipment means that the instruments used by engineers in research, development, and product testing must be of an ever improving quality and versatility. To keep instrumentation costs within reasonable bounds, and simultaneously produce a high quality versatile instrument, a number of instrument manufacturers are now marketing instrumentation systems which are eminently suitable for this type of situation. Some examples are: (a) the Tektronix 7000 series oscilloscope[3] (see figure 2.14) which in addition to the normal oscilloscope facilities this instrument may be fitted with additional plug-in modules so that measurements of time interval, frequency and magnitude of the measurand may be displayed in numerical form alongside the trace of the waveform — these facilities having now been further enhanced by the addition of a processor[9] that can be used to digitise the acquired signal, provide storage and perform computation; and (b) the Hewlett Packard 3480 A, multifunction meter[4] which with the aid of plug-in units may be used for measuring such quantities as direct voltage, voltage ratios, alternating voltage, and resistances. A number of such systems[10] are appearing on the market, so that a survey of manufacturers' literature is the only method of obtaining a satisfactory knowledge of the 'state of the art', and in conducting such a survey the factors discussed in chapter 10 should be carefully observed. It is possible that the instrumentation for the development laboratory bench of the future will tend to incorporate such a system in that these are really updated versions of the well tried multimeter. For production situations, the data logger or computer controlled type of system is more likely to be the norm of the future since these incorporate record keeping with the programmable features.

11.3 PERMANENT INSTALLATIONS

The automatic instrumentation system which scans the outputs of a number of sensors, compares their magnitudes with predetermined reference levels, and records the measurands in a programmed manner without the assistance of an attendant has been developed form the completely manual system that required a number of attendants to tour the installation and report at regular intervals to a controller. An intermediate stage in the development is represented by the semi-automatic system, incorporating both local and remote indication of measurands, and possibly including such devices as pen recorders and audible alarms.

To illustrate types of permanent instrument installations two examples are given. However, it should be appreciated that there are a number of satisfactory solutions for every problem of this type, and each solution will be influenced by an engineer's experience in the particular field and his personal preference for one type of instrument.

Example A

It is required to outline an instrumentation system capable of monitoring the performance of an oil burning 50 MW power station to ensure safe and efficient operation.

The solution given is not intended to be a specification for an instrumentation system, but to indicate the differences between manual, semiautomatic, and automatic systems, and show how a problem of this nature may be tackled. A convenient method of commencing on the solution is to tabulate the measurands and compare the methods by which they could be measured (see table 11.1).

The prime purpose of any of these systems must be to ensure efficient operation and prevent serious damage in the event of a fault developing.

Manual system

The system requires attendants to tour the station at regular intervals (say half hourly) and record readings on all the meters and indicators, and then report to the controller. The latter should have at least a mimic display to monitor the important quantities such as drum water level, boiler temperatures and pressures, and alternator output, the rounds of the attendant merely serving to ensure that the conditions existing in the actual plant are faithfully reproduced by the mimic displays on the controller's panel. To record the performance of the station, attendants must tabulate and log all quantities by hand, or no permanent record will be made.

Semiautomatic system

In using a semiautomatic system of instrumentation a considerable saving of

Measurand	Manual system	Semiautomatic	Automatic
Boiler			
—fuel flow	flow meter	flow meter	transducer
—inlet air flow and temperature	flow meter: thermometer	flow meter: thermocouple	transducer: thermocouple
—temperatures at each stage	thermometers	thermometers or thermocouples	thermocouples
—pressures	bourdon tube gauge	pressure gauges	pressure transducers
—drum water level	level gauge	level gauge, closed circuit TV link	level transducer, level gauge and closed circuit TV
—feed water	thermometer: flow meter	thermocouple: flowmeter	thermocouple: flow sensor
Turbine			
—stage pressures			
—stage temperatures	thermometers in thermometer pockets bourdon tube pressure gauges audible alarms	thermocouples and pen recorders for temperature measurement pressures either on chart recorders or read at intervals	thermocouples, pressure transducers, level sensors, to data logger
—condensate temps			
—condensate pressures			
—bearing pressures			
—bearing temps			
—speed			
Alternator			
—output voltage	VT plus meter	instrument transformers plus meters	instrument transformers plus meters and conditioning circuits*
—output current	CT plus meter		
—VA ouptut			
—power output	instrument transformers plus meters		
—phase angle			
—field current			
—frequency			
—rotor temperatures			
—stator temperatures	thermometers	thermocouples	thermocouples
—bearing temperatures			
—leakage currents	CT plus alarms	CT plus alarms	CT plus alarms
Accessories			
—pumps functioning			
—lubrication	indicator plus alarm	indicator plus alarm	indicator plus alarm
—cooling water			

* For example cooling circuits pages 153, 161, and lingerisers page 277

labour is achieved by the use of pen recorders to log temperatures, and possibly some of the pressures, by using either single or multichannel recorders, possibly fitted with alarm switches that are operated if a predetermined level is exceeded. The instruments for displaying and recording the measurands would almost certainly be positioned in the control room; the appropriate quantities may also be displayed in the boiler and turbine room, these being linked by direct telephone to the control room. Because of the importance of the water level in the boiler drum the expense of displaying this level in the control room by using a closed circuit TV link may be justified.

Automatic system

An automatic system is most likely to take the form of some type of data logging system. This enables all the variables to be scanned at a moderately fast rate (say 100/s) to check that limit conditions are not exceeded, the readings of the measurands would be logged in tabular form at regular intervals, for example hourly. Such a system may be linked to a computer programmed to control the installation so that steam temperatures and pressures, fuel feed rates, alternator field current, etc., are continually adjusted to maintain a condition of optimum efficiency in the working of the station. It is, however, likely that all the measurands will also be monitored by analog instruments on the control engineer's panel so that he is still in touch with the performance of the station and is capable of quickly taking control should a fault develop in the automatic recording/control system.

Example B

This example examines the instrumentation associated with the testing on a production line of a 'plug-in' power amplifier.

Testing such a device will require the application of power supplies to energise it, the measurement of its response, for example bandwidth, gain, phase shift, temperature rise, drift, input and output impedances, and determining the effects of variations in the load and the supply voltage. The resulting readings must be tabulated and compared with the performance as quoted in the amplifier specification.

As with example A there are three methods of tackling the problem:

(a) *Manual.* This is cheap in equipment but expensive in labour costs, requiring a test engineer to perform a routine of tests and procedures to establish the performance of the amplifier. Such a system will inevitably result in the operator taking short cuts if a large number of similar units are to be tested, for he will know what to expect at each stage and may be tempted to look for these values rather than recording the magnitudes that actually occur.

(b) *Semiautomatic.* In this type of system the operator of system (a) is assisted by instruments that remove much of the monotony. For example, a sweep oscillator and X–Y recorder could be used to plot the frequency response of the amplifier; a multipoint pen recorder could be used to plot the amplifier input and output voltages over a cycle of supply voltage and load variations; similarly, the temperatures at various points in the amplifier could be monitored during an elongated proving test by using thermocouples and a potentiometric recorder.

(c) *Automatic.* To make the system automatic requires the inclusion of a programmed sequence control device[5]. This may be a set of cam switches, or a sequence programmer; if the operation is more complex the programme of tests requiring variations in supplies, signals, environment, etc., may be controlled by a programme stored on magnetic tape or controlled by a mini computer[6,7]. The effects of various levels of signal applied at points throughout the circuit under test are recorded by printer or on magnetic tape or punched tape. Such a system will perform tests much more rigidly that a human operator and enables better quality control and cyclic testing to be performed at more stages in a manufacturing process.

REFERENCES

1 *Encyclopedia of Science and Technology,* McGraw-Hill, New York (1971)
2 G. A. T. Burdett. *Automatic Control Handbook,* Newnes, London (1962)
3 *7000 Series Oscilloscopes and Accessories,* Tektronix, Beaverton (1971)
4 C. Walter, H. MacJuneau and L. Thompson. 'A New High Speed Multi-function Digital Voltmeter'. *Hewlett Packard Journal* (Jan. 1971)
5 O. G. Pamely-Evans. 'Techniques Used in Programme and Sequence Control'. *Control and Instrumentation.* 35–37 (June 1971)
6 R. A. Grim. 'Automated Testing'. *Hewlett Packard Journal* (Aug. 1969)
7 'Automated Measurement Systems'. *Tekscope,* 2, No. 4, Tektronix (1970).
8 D. M. Considine. *Encyclopedia of Instrumentation and Control,* McGraw-Hill, New York (1971)
9 'The Oscilloscope with Computing Power', *Tekscope,* 5, No. 2, Tektronix (1973)
10   'Test and Measurement System 500', Tektronix Beaverton (1973)

Appendix I

The international system of units (Système International d'Unités – SI) has been established by international agreement to provide a logical and interconnected framework for all measurements in science and industry. It is based on six units: the metre (length), the second (time), the kilogramme (mass), the ampere (electric current), the kelvin (temperature) and the candela (luminous intensity). The majority of the world's population already use the SI system of units and it is possible that by 1975 the only major countries who have not changed will be Canada and the U.S.A.

In addition to the defining of the six basic units the SI establishes a set of factors that should be applied to all quantities so that their magnitude may always be written with less than four numerals before the decimal point, thus removing some of the difficulties associated with manipulating very large and very small numbers (see table A.1). Table A.2 lists the units in the SI system.

Table A.1 Multiples

Prefix	Symbol	Factor by which the unit is to be multiplied
tera	T	10^{12}
giga	G	10^{9}
mega	M	10^{6}
kilo	k	10^{3}
milli	m	10^{-3}
micro	μ	10^{-6}
nano	n	10^{-9}
pico	p	10^{-12}
femto	f	10^{-15}
atto	a	10^{-18}

Table A.2. Units

Quantity	SI Units	Standard – definition	Equivalents Imperial	Equivalents M.K.S.	Conversion factor
Mass	kilogramme (kg)	mass of a platinum-iridium cyclinder kept at Sevres, France	pound (lb)	(kg)	1 lb = 0.454 kg 1 cwt = 50.8 kg 1 ton = 1016 kg
Length	metre (m)	distance between two engraved lines on a platinum-iridium bar kept at Sevres, France – also as 1 650 763.73 wavelengths in vacuo of the orange line (spectroscopic designation $2\,p_{10}\,5\,d_5$) emitted by the krypton-86 atom	foot (ft)	(m)	1 in = 0.0254 m 1 ft = 0.3048 m 1 yd = 0.9144 m 1 mile = 1.61 km
Time	second (s)	the fraction 1/31 556 925.975 of the tropical year for 1900 Jan. 0. Also as the interval occupied by 9 192 631 770 cycles of the radiation corresponding to the transition between the two hyperfine levels of the ground state of caesium 133	(s)	(s)	
Temperature (absolute)	kelvin (K)	the degree interval on the thermodynamic scale on which the temperature of the triple point of water is 273.16 K	degree fahrenheit (°F)	degree centigrade (°C)	
(interval)	Celsius (°C)	temperature difference in degrees Celsius (≡ Centigrade) is the same as in Kelvins	(°F)	(°C)	1°F = 5/9°C
(scale)	Celsius (°C)	temperature scale in degrees Celsius is the scale value in Kelvin minus 273.15	(°F)	(°C)	0°C = 273.15 K

Quantity	SI unit	Definition	Other units	Conversions
Electric current	ampère	the current that if flowing in two infinitely long parallel wires spaced 1 m apart in vacuo, would produce a force of 2×10^{-7} N per metre of length between the wires	A	
Force	newton (N) N = kg m/s² N = J/m	that force which, when acting on a mass of 1 kg gives it an acceleration of 1 m/s²	kilogramme (kgf) pound (lbf)	1 kgf = 9.8066 N 1 lbf = 4.4482 N
Work, energy, quantity of heat	joule (J) J = N m J = W s	the work done by a force of 1 N when its point of application is moved through a distance of 1 m in the direction of the force. This unit is used for every kind of energy including heat	m kgf ft lbf	1 ft lbf = 1.3558 J 1 m kgf = 9.8066 J 1 eV = 1.6021×10^{-19} J 1 cal = 4.1868 J 1 Btu = 1055.06 J
Power	watt (W) W = J/s = N m/s	1 J/s	Watt (W)	1 hp = 745.7 W
Electric charge	coulomb (C) C = A s	the quantity of electricity transported in 1 s by a current of 1 A	Coulomb (C)	
Electric potential	volt (V) V = W/A	the difference of potential between two points of a conducting wire which carries a constant current of 1 A when the power dissipated between these two points is 1 W, or 1 J/s	Volt (V)	
Electric capacitance	farad (F) F = A s/V F = C/V	a capacitance is 1 F if a difference of potential of 1 V appears between the plates of a capacitor when it is charged with 1 C of electricity	farad (F)	

Table A.2 (cont.)

Quantity	SI Units	Standard – definition	Equivalents Imperial	M.K.S.	Conversion factor
Electric resistance	ohm (Ω) $\Omega = V/A$	the resistance between two points of a conductor when a constant difference of potential of 1 V applied between these two points produces a current of 1 A in the conductor. The conductor must not be the source of any electro-motive force at that time	ohm (Ω)		
Magnetic flux	weber (Wb) Wb = V s	the flux which, when linking a circuit of one turn, and being reduced to zero at a uniform rate in one second produces in the circuit an electromotive force of 1 V	weber (Wb)		
Magnetic flux density	tesla (T) $T = Wb/m^2$	Magnetic flux per square metre		Wb/m^2	1 gauss = 10^{-4} C
Electric inductance	henry (H) H = V s/A H = Wb/A	the henry is the inductance of a closed circuit in which an electro-motive force of 1 V is produced when the electric current in the circuit varies uniformly at the rate of 1 A/s	henry (H)		
Electric conductance	siemens (S) S = A/V S = 1/Ω	reciprocal of resistance	mho ($\mho$)		
frequency	hertz (Hz) Hz = 1/s	number of complete periods of oscillation per second	cycles per second (C/S)		

For fuller details see:

1. *Symbols and Abbreviations for use in Electrical and Electronic Engineering Courses.* Published by the Institution of Electrical Engineers (1968).
2. B.S. 1991, *Letter Symbols, Signs and Abbreviations.*
3. B.S. 3763, *International System (SI) Units.*
4. PD 5686, The Use of SI Units, BSI (1972).

Appendix II

DYNAMIC BEHAVIOUR OF MOVING COIL SYSTEMS

a. Equation of motion; damping magnitude

It is indicated in chapter 1 that a number of factors affect the movement of the coil of an instrument, these being:

    (a)   the moment of inertia J
    (b)   the damping constant D
    (c)   the control constant or spring stiffness C

These factors may be equated with the deflecting torque (GI) to form the equation of motion

$$J\frac{d^2\theta}{dt^2} + D\frac{d\theta}{dt} + C\theta = GI \tag{II.1}$$

To obtain the transient solution of this equation consider the deflecting torque (GI) to be removed; the moving coil then twisted through an angle and released. The equation of motion then becomes:

$$J\frac{d^2\theta}{dt^2} + D\frac{d\theta}{dt} + C\theta = 0 \tag{II.2}$$

This equation is satisfied by a solution of the form

$$\theta = k_1 e^{\lambda t} \tag{II.3}$$

differentiating equation II.3 gives

$$\frac{d\theta}{dt} = k_1 \lambda e^{\lambda t} \quad \text{and} \quad \frac{d^2\theta}{dt^2} = k_1 \lambda^2 e^{\lambda t}$$

Substituting in equation (II.2) gives

$$J\lambda^2 e^{\lambda t} + D\lambda e^{\lambda t} + Ce^{\lambda t} = 0 \tag{II.4}$$

$$\text{or} \qquad J\lambda^2 + D\lambda + C = 0 \tag{II.5}$$

if λ_1 and λ_2 are the two roots of this equation

$$\lambda_1 = \frac{-D + (D^2 - 4CJ)^{\frac{1}{2}}}{2J}$$

and $$\lambda_2 = \frac{-D - (D^2 - 4CJ)^{\frac{1}{2}}}{2J}$$

As two arbitary constants are required in the solution of a second order differential equation, the complete solution of (II.2) is

$$\theta = Ae^{\lambda_1 t} + Be^{\lambda_2 t} \tag{II.6}$$

where A and B are the arbitrary constants, the values of which may be obtained from the conditions of motion.

It is necessary to consider three cases:

(1) Overdamped

$D^2 > 4CJ$ – the roots are real and unequal, the solution taking the form of the sum of two quantities both of which diminish exponentially.

that is $$\theta = \frac{\theta_0}{\lambda_2 - \lambda_1} \left\{ \lambda_2 e^{\lambda_1 t} - \lambda_1 e^{\lambda_2 t} \right\} \tag{II.7}$$

being the curve (a) in figure A.1.

(2) Critically damped

$D^2 = 4CJ$ – the roots being equal and real. The curve (b) in figure A.1 being computed from the equation

$$\theta = \theta_0 \cdot \left(1 + \frac{D}{2J} \times t \right) e^{\lambda t} \tag{II.8}$$

(where $\lambda = -D/2J$). This is a special case, being termed critical damping, and is the condition in which an instrument coil will change from one position to another in a minimum of time without overshoot.

(3) Underdamped

$D^2 < 4CJ$ – the roots becoming conjugate – complex

$$\lambda_1 = -\alpha + j\omega \quad \text{and} \quad \lambda_2 = -\alpha - j\omega$$

$$\alpha = \frac{D}{2J} \quad \text{and} \quad \omega = \left(\frac{C}{J} - \alpha^2 \right)^{\frac{1}{2}}$$

Substituting these values of λ_1 and λ_2 in II.6 gives

$$\theta = e^{-\alpha t} \left\{ Ae^{j\omega t} + Be^{-j\omega t} \right\}$$

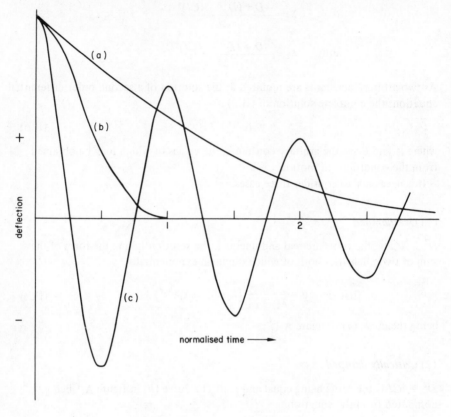

Figure A.1. Dynamic behaviour of a moving coil instrument

Since $e^{\pm j\omega t}$ is complex and θ must be real as it represents a physical quantity, it follows that A and B must be complex.

Let $A = a + jb$ and $B = c + jd$, then if $e^{\pm j\omega t} = \cos \omega t \pm j \sin \omega t$

$$\theta = e^{-\alpha t} \left\{ (a + jb) \left[\cos \omega t + j \sin \omega t \right] + (c + d) \left[\cos \omega t - j \sin \omega t \right] \right\}$$

and since the imaginary part must be zero for all values of t

$$(b + d) \cos \omega t + (a - c) \sin \omega t = 0$$

now $a = c$ and $b = -d$ hence A and B must be complex conjugates.
The real part of θ which remains is

$$\theta = 2e^{-\alpha t} (c \cos \omega t + d \sin \omega t) \tag{II.9}$$

which may be expressed as a single trigonometric function as

$$\theta = U e^{-\alpha t} \sin (\omega t + \phi)$$

where $U = 2(c^2 + d^2)^{\frac{1}{2}}$ and $\phi = \tan^{-1} \dfrac{c}{d}$

Showing that the transient is a damped sinusoid, of angular frequency

$$\omega = \left[\frac{C}{J} - \left(\frac{D}{2J} \right)^2 \right]^{\frac{1}{2}}$$

as shown by curve (c) in figure A.1.

If the instrument coil has only a small amount of damping so that $D^2/4J^2$ may be neglected in comparison with C/J

$$\omega_0 = \left(\frac{C}{J} \right)^{\frac{1}{2}} \tag{II.10}$$

giving the natural frequency of oscillation of the instrument coil as

$$f_0 = \frac{1}{2\pi} \left(\frac{C}{J} \right)^{\frac{1}{2}} \tag{II.11}$$

b. Deflection amplitude of vibration galvanometer

The vibration galvanometer (see chapter 1) has a narrow air cored coil of low inertia, and a stiff suspension (large control constant). If the current in such a coil is $i = I_m \sin \omega t$ the equation of motion is

$$J \frac{d^2\theta}{dt^2} + D \frac{d\theta}{dt} + C\theta = GI_m \sin \omega t \tag{II.12}$$

In this case the transient solution is not of practical importance but the steady state solution will be and is obtained from the particular integral which is of the form

$$\theta = A \sin (\omega t - \phi) \tag{II.13}$$

where A and ϕ are constants.

Now $\dfrac{d\theta}{dt} = A\omega \cos (\omega t - \phi)$ $\tag{II.14}$

and $\dfrac{d^2\theta}{dt^2} = -A\omega^2 \sin (\omega t - \phi)$ $\tag{II.15}$

Substituting (II.13, II.14 and II.15) in II.12 gives

$$-AJ\omega^2 \sin (\omega t - \phi) + A D\omega \cos (\omega t - \phi) + AC \sin (\omega t - \phi) = GI_m \sin \omega t \tag{II.16}$$

This expression must be true for all values of t and when $\omega t = \phi$

$$A D\omega = GI_m \sin \phi \tag{II.17}$$

and when $(\omega t - \phi) = \pi/2$

$$-AJ\omega^2 + AC = GI_m \cos\phi \qquad (II.18)$$

As the phase angle ϕ is of no practical significance — the amplitude of movement being the important quality when considering a vibration galvanometer, ϕ may be eliminated by squaring and adding equations II.17 and II.8.
Hence

$$A^2D^2\omega^2 + A^2(C - J\omega^2)^2 = G^2I_m^2$$

or $\qquad A = \dfrac{GI_m}{\left[D^2\omega^2 + (C - J\omega^2)^2\right]^{\frac{1}{2}}} \qquad (II.19)$

A being the amplitude of the oscillation resulting from the application to the coil of an alternating current having a peak value of I_m.

Now for a given frequency ω, amplitude will be a maximum if $\omega = (C/J)^{\frac{1}{2}}$ which is also the angular frequency of undamped oscillation of the moving coil (equation II.11). Thus a vibration galvanometer should be 'tuned' so that the frequency of its undamped oscillations is equal to the system frequency. The amplitude of oscillations for other frequencies being calculated using equation II.19.

c. Amplitude and phase distortion of a u.v. recorder galvanometer

Consider a current $I_m \sin \omega t$ applied to a recorder galvanometer. Since the construction of a recorder galvanometer (see page 51) is similar to that of a vibration galvanometer in that it has low inertia, is air cored, and has a moderately stiff suspension the amplitude of vibrations will be described by the equation II.19, namely

$$A = \frac{GI_m}{\left[D^2\omega^2 + (C - J\omega^2)^2\right]^{\frac{1}{2}}}$$

Now an 'ideal' recorder galvanometer would have zero damping and inertia effects, that is it would respond exactly to any current variations whatever the frequency. In such a case the amplitude A_i (ideal) would be GI_m/C (for $D = 0$ and $J = 0$). Now in the u.v. recorder, resonance effects and troubles due to transients are reduced by damping of the galvanometer. Critical damping would lead to a certain amount of distortion but by using less than critical damping an optimum of performance may be obtained (see chapter 2).

If a damping (d) is used such that $d = \eta D$ [D being critical damping $= (4CJ)^{\frac{1}{2}}$]

The amplitude A_a (actual) $= \dfrac{GI_m}{\left[d^2\omega^2 + (C - J\omega^2)^2\right]^{\frac{1}{2}}} \qquad (II.20)$

$$= \frac{GI_m}{(\eta^2 D^2 \omega^2 + C^2 - 2J\omega^2 + J^2\omega^4)^{\frac{1}{2}}}$$

now $\quad D = (4CJ)^{\frac{1}{2}}; \quad$ and $\quad \dfrac{J}{C} = \dfrac{1}{\omega_0{}^2}$

$$A_a = \frac{GI_m}{C\left(\eta^2\, 4\dfrac{\omega^2}{\omega_0{}^2} + 1 - 2\dfrac{\omega^2}{\omega_0{}^2} + \dfrac{\omega^4}{\omega_0{}^4}\right)^{\frac{1}{2}}}$$

putting $\omega_r = \omega/\omega_0$

$$A_a = \frac{GI_m}{C(1 - 2\omega_r^2 + \omega_r^2 + 4\eta^2\omega_r^2)^{\frac{1}{2}}}$$

or $\quad A_a = \dfrac{GI_m}{C\left[(1 - \omega_r^2)^2 + (2\eta\omega_r)^2\right]^{\frac{1}{2}}}$ $\qquad$ (II.21)

$\therefore$ Relative amplitude of A_a to A_i is

$$\frac{A_a}{A_i} = \frac{1}{\left[(1 - \omega_r^2)^2 + (2\eta\omega_r)^2\right]^{\frac{1}{2}}} \qquad (\text{II.22})$$

Now $\omega_r = \omega/\omega_0 = f/f_0 = f_r$ the ratio of the frequency of the signal to the resonate frequency of the galvanometer.

$$\therefore \quad \frac{A_a}{A_i} = \frac{1}{\left[(1 - f_r^2)^2 + (2\eta f_r)^2\right]^{\frac{1}{2}}} \qquad (\text{II.23})$$

To estimate the amplitude distortion in any recorded wave it is necessary to calculate A_a/A_i for the fundamental and for each harmonic.

To assess the possible phase distortion of harmonics or the time displacement error between channels using different galvanometers it is necessary to reconsider equations II.17 and II.18 from which:

$$\tan\phi = \frac{G \cdot I_m \sin\phi}{G \cdot I_m \cos\phi}$$

$$= \frac{A\, d\omega}{AC - AJ\omega^2}$$

$$= \frac{\eta\,(4CJ)^{\frac{1}{2}}\,\omega}{C - J\omega^2}$$

$$= \frac{2\eta\omega}{\left(\dfrac{C}{J}\right)^{\frac{1}{2}} - \left(\dfrac{J}{C}\right)^{\frac{1}{2}}\omega} = \frac{2}{\dfrac{\omega_0}{\omega} - \dfrac{\omega}{\omega_0}}$$

$$\therefore \quad \tan\phi = \frac{2\eta}{\dfrac{1}{f_r} - f_r} = \frac{2\eta f_r}{1 - f_r^2} \qquad (\text{II.24})$$

Note. This expression gives the phase displacement of the recorded frequency with respect to its true position, *measured on its own time scale* of angular frequency. To estimate the distortion produced in a recorded wave it is necessary to calculate the phase displacement of the fundamental and each harmonic, and to allow for the different angular frequency scales of the various harmonics.
For example a 25° displacement of a 5th harmonic on its own time scale represents 5° displacement on the angular frequency scale of the fundamental or vice versa.

For further analytical work see:

E. Frank. *Electrical Measurement Analysis,* McGraw-Hill, New York (1959).

Appendix III

**EQUATIONS TO DETERMINE THE COMPONENTS OF A RESISTIVE 'T'
ATTENUATOR PAD**

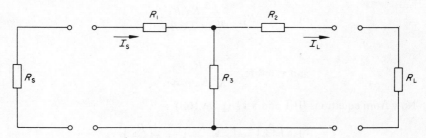

Figure A.2. Resistive attenuator pad

Consider the resistive attenuator pad shown in figure A.2 for which

$$\frac{I_s}{I_L} = \frac{R_3 + R_2 + R_L}{R_3} = K \qquad (III.1)$$

For correct matching (see page 150)

$$R_s = R_1 + \frac{R_3 (R_2 + R_L)}{R_2 + R_3 + R_L} \qquad (III.2)$$

and

$$R_L = R_2 + \frac{R_3 (R_1 + R_s)}{R_1 + R_3 + R_s} \qquad (III.3)$$

From equation III.1

$$\frac{R_2 + R_L}{R_3} = K - 1$$

or

$$R_2 + R_L = R_3 (K - 1) \qquad (III.4)$$

now from equations III.1 and III.2

$$R_1 + \frac{(R_2 + R_L)}{K} = R_s \qquad (III.5)$$

Substituting equation III.4 in III.5 gives:

$$R_1 + \frac{R_3(K-1)}{K} = R_s$$

or

$$R_1 = \frac{R_3(1-K)}{K} + R_s \tag{III.6}$$

Now substituting equations III.4 and III.6 in III.3 gives:

$$2R_L = R_3 \left\{ (K-1) + \frac{R_3\left(\frac{1}{K}-1\right) + 2R_s}{R_3\left(\frac{1}{K}-1\right) + 2R_s + R_3} \right\}$$

or

$$R_L = \frac{K^2 R_3 R_s}{R_3 + 2K R_3}$$

and

$$R_3 = \frac{2K R_s R_L}{K^2 R_s - R_L} \tag{III.7}$$

Now from equations III.1 and 5.11 (page 168)

$$k = \left(\frac{P_s}{P_L}\right)^{\frac{1}{2}} = \left(\frac{I_s^2 R_s}{I_L^2 R_L}\right)^{\frac{1}{2}} = K\left(\frac{R_s}{R_L}\right)^{\frac{1}{2}}$$

or

$$K = \frac{k(R_L R_s)^{\frac{1}{2}}}{R_s} \tag{III.8}$$

Substituting equation III.8 in III.7 gives

$$R_3 = \frac{2k(R_s R_L)^{\frac{1}{2}}}{k^2 - 1} \tag{III.9}$$

Substituting equations III.8 and III.9 in III.4 gives

$$R_2 = R_L\left[\frac{k^2+1}{k^2-1}\right] - \frac{2k(R_s R_L)^{\frac{1}{2}}}{k^2-1} \tag{III.10}$$

and substituting equations III.8 and III.9 in III.6 gives

$$R_1 = R_s\left[\frac{k^2+1}{k^2-1}\right] - \frac{2k(R_s R_L)^{\frac{1}{2}}}{k^2-1} \tag{III.11}$$

Note. $k = (P_s/P_L)^{\frac{1}{2}}$; and attenuation, $A = 10 \log_{10}(P_s/P_L)$ dB

Hence $k = 10^{A/20}$ or $e^{0.11513A}$ where A is in dB.

Index